Fundamentals of Integrated Vehicle Realization

Other SAE books of interest:

The Evolution of Automotive Technology: A Handbook
By Gijs Mom
(Product Code: R-435)

Project Management for Automotive Engineers: A Field Guide
By Jon M. Quigley and Roopa J. Shenoy
(Product Code: R-437)

Software-Hardware Integration in Automotive Product Development
By John Blyler
(Product Code: PT-161)

Fundamentals of Integrated Vehicle Realization

Mohamed El-Sayed, P.E.

SAE INTERNATIONAL®

Warrendale, Pennsylvania, USA

INTERNATIONAL®

400 Commonwealth Drive
Warrendale, PA 15096-0001 USA
E-mail: CustomerService@sae.org
Phone: 877-606-7323 (inside USA and Canada)
724-776-4970 (outside USA)
Fax: 724-776-0790

SAE Order Number R-436
http://dx.doi.org/10.4271/r-436

Library of Congress Control Number: 2016949681

ISBN-Print 978-0-7680-8036-0
ISBN-PDF 978-0-7680-8383-5
ISBN-epub 978-0-7680-8385-9
ISBN-prc 978-0-7680-8384-2

To purchase bulk quantities, please contact:

SAE Customer Service
E-mail: CustomerService@sae.org
Phone: 877-606-7323 *(inside USA and Canada)*
724-776-4970 *(outside USA)*
Fax: 724-776-0790

Visit the SAE International Bookstore at books.sae.org

Contents

Preface

Continuous demand for product refinement and added functionality is one of the most persistent issues facing many organizations in highly competitive markets. As products become more refined and complex, new specialized knowledge, skills, and expertise are needed. Facing this challenge, organizations are under pressure to continuously transform themselves. As a result, many rapidly developed organizational structures are fashioned and implemented. In most cases, however, these reorganization efforts do not improve product development as planned or desired. In addition to organizational instability due to fluidity during any rapid transformation and the need for adjustment, most of the resulting structures lack the natural alignment with the product development phases. Consequently, many product development initiatives may be canceled or dropped at later phases despite all preceding efforts and financial investments. Therefore, it is vital that organizational unity and harmony be always intact during any transformation. A highly effective organization should always act as one cohesive entity dedicated to serving the customer with creative aptitude, integrative skills, analytical thinking, and synergistic management. To these organizations and individuals, continuously striving to make life better for everyone, I dedicate this book.

My utmost praise and gratitude is to the Creator for all the bounties bestowed upon me. To my wife, Dr. Jacqueline El-Sayed, is my sincere appreciation for her encouragement, patience, and support. My gratitude is for all the help she provided to me during the course of writing this book, for always being my best friend, cheering me up when life got tough, and for advising me when I needed. Many thanks are also to my children and extended family for acting as sounding boards, sharing their thoughts, and providing insights to many of the new realization concepts discussed in the book.

I would like to acknowledge and thank Dr. Steve Czarnecki for the many discussions we had and the reviews he had provided. For instance, in addition to many of his valuable insights, he pointed out the repeated or fractal-like pattern of the realization phases and subphases. I would also like to acknowledge and thank Dr. Marcus Cappelli and Dr. Steve Beyerlein for their comprehensive reviews and edits. Thanks are also due to all the mentors, managers, engineers, professors, teachers, and students who helped me grow my knowledge and develop my skills throughout my career.

Finally, I would like to thank and acknowledge Monica Nogueira and Mandy May from SAE for their valued help and guidance provided to me. I also would like to thank the

SAE organization, leadership, staff, and members for the continual support to me and the mobility industry.

I.1 Introduction

The term product realization is used by the automotive quality management system international standard ISO/TS 16949:2002 [I-1], [I-2], "to describe the work that the organization goes through to develop, manufacture, and deliver the finished goods or services." In addition, product realization as a process is epitomized through the adoption of the Advanced Product Quality Planning [3] as the product realization process for the standard. With this usage of the term, it is clear that product realization encompasses other work and activities with product development. However, the full extent of realization as a concept or a process is not well established, understood, or explored. From the current literature and practices, very few and limited efforts have been directed toward the underlying fundamentals of the realization process [I-4], [1-5], [I-6]. Most the efforts are focused on product development or process management. In an effort to fill the current void, this book was written to address the core concept of realization in the context of bringing products and processes to reality. By studying the roots of the realization concept, the main objective is to establish the foundation and demonstrate the implementation of integrated product realization process thinking and practices.

Integrated realization process thinking and implementation require balanced, comprehensive, and holistic views of the process from the perspective of product development as well as process management. One of the main issues that can be easily observed from current practices and the existing literature is the lack of integration between technical product development and process thinking knowledge. While technical literature, research efforts, and practices are focused on the product and its development challenges, process research efforts and literature are mainly focused on the process flow and its management. Managing the process without understanding the product and its development challenges could cause unrealistic expectations, erroneous decision making, and unintegrated realization efforts. On the other hand, carrying out product development without understanding process flow, timing, and assigned roles could adversely affect product quality, time to market, and cost. Therefore, seamless integration of realization efforts, in any organization, starts with management that has a deep understanding of the product development challenges and a complementary technical team that has a high level of comprehension of the development process nuances. To bridge these product development and process management gaps, this book considers both sides by addressing product realization from a process perspective and realization process thinking from the product development viewpoint.

Effective and lasting integration in product realization would not be achieved without challenging some of the prevalent paradigms and associated practices. To alter these paradigms and stimulate new practices, a fresh product realization paradigm, along with holistic integrated product and process realization frameworks, is presented.

To demonstrate the implementation of the integrated realization concept and the holistic frameworks, integrated vehicle realization as one of the most demanding and comprehensive processes is addressed. Since process thinking, mass production, and the assembly line concept are deeply rooted in the automotive industry and vehicle production, addressing vehicle realization demonstrates most of the realization process elements needed for the realization of other products. Accordingly, the presented realization approach is applicable to product realization in any context.

The approach followed in this book is to take a fresh look and rethink the realization process starting with the realization concept itself. To help in the logical construction of the presented knowledge, the book builds on the historical progression, best practices, and the observed principles of human realization. Accordingly, the realization concept and holistic frameworks presented are founded on observations of deeply rooted principles and centuries of product development practices. To facilitate the grasp of these newly introduced realization concepts and frameworks, the book is deliberately focused on the fundamentals.

Since product realization covers all the work that an organization goes through, different levels of technical and management knowledge and skills are anticipated from the individuals involved. To benefit most of these individuals, *Fundamentals of Integrated Vehicle Realization* is designed to be accessible to readers with diverse backgrounds and experience levels. In addition to practicing engineers, managers, entrepreneurs, and corporate leaders, the book can be adopted as a text or a reference for students in higher education. The book can also be used in graduate and upper-level undergraduate courses in the fields of product design, capstone design, product development, system engineering, engineering management, process development, and process management.

Fundamentals of Integrated Vehicle Realization is organized in logically connected building blocks to help different readers in constructing their own learning. Each building block has the necessary and sufficient depth, case studies, and examples for a learning experience without overwhelming details. However, additional references are provided for further study of the "how-to" aspects to fit the needs of different readers. The two foundational building blocks of historical progression of process evolution and the realization concept are deliberated in Chapters 1 and 2. The two essential building blocks of process and product realization frameworks are exemplified in Chapters 3 and 4. The two key desirable and most challenging drivers behind the integrated realization framework—namely, quality and innovation—are discussed in Chapters 5 and 6. The three structured realization phases and their application in the context of vehicle realization are illustrated in Chapters 7–9. Finally, some of the key enablers for successful implementation of the integrated product realization concept and frameworks are encapsulated in Chapter 10.

Each chapter is designed to facilitate learning of the introduced holistic and integrated realization concept and frameworks. To demonstrate the concepts and explain the relationships being described, each chapter contains some illustrative examples, case studies, and original color-coded graphics. Some of them are repeated, detailed, or expanded in later chapters to aid in complex concept building and in clarifying different relationships. To help different readers in grasping and implementing the frameworks and realization concepts discussed, each chapter is concluded by the desired learning outcomes in terms of knowledge summary, expected skills, and anticipated paradigm shifts. A brief summary for the aim and scope of each chapter is presented below.

Due to the extreme importance of history in understanding the present and building on the lessons learned, the evolution of the vehicle development process is studied in Chapter 1. The main objective for this study is better understanding of the current state of product development process and the associated issues. The study, however, did not focus only on the historical progression, but also on the in-depth analyses of the different development process models. The aim of these analyses is to decipher and understand the factors shaping and driving the development of the evolving process models. Understanding the factors behind the different process models is key to understanding the state of the art in process development and how it was arrived at. In addition, these factors are necessary for the development of the holistic product realization framework presented in the succeeding chapters.

The construction of the realization process introduced in this book, by taking a different look at the concept of realization, is discussed in Chapter 2. The introduced realization concept and its associated terminologies are essential for the development of a new and much needed product realization paradigm and a well-integrated realization process. The main objective of the chapter and its discussions is to better understand the introduced realization concept in the context of product, process, and development. Instead of starting the chapter with unfamiliar terminologies and definitions, the discussion provided aims at building the concept through the expansion of the current use and understanding of the word *realization*.

In Chapter 3, a holistic framework for process characteristics, attributes, and criteria is introduced. Process characteristics and attributes are classified and the associated sets of criteria shaping each attribute are identified. The development of this framework is illustrated by some examples of desired process characteristics, attributes, and criteria. To demonstrate the implementation of the framework and the achievement of the aspired process, samples of process targets for different criteria are also discussed.

A key step in integrated vehicle realization is discussed in Chapter 4. The holistic framework put forth in Chapter 3 is applied to vehicle realization. Consequently, vehicle characteristics and attributes are classified, and samples of the associated sets of criteria shaping each attribute are identified. Some examples of vehicle-level targets, for different criteria, are discussed to illustrate the use of the framework.

Quality as a central element of holistic realization frameworks and as a fundamental driver for integrated product realization is studied in Chapter 5. In-depth analyses are conducted for quality and its relationship to the gaps in product characteristics and attributes. Consequently, potential losses due to these gaps are examined. In addition, achievements of product quality and production process improvability are studied. Process manageability and improvability for achieving product quality during product realization are discussed, along with some examples of quality tools and methodologies.

The front-end activities and the fuzzy end of innovation as integral parts of product realization are discussed in Chapter 6. In addition to the new front-end innovation model presented, activities such as screening, idea generation, and concept evaluation are studied. The new front-end innovation model proposes balancing the two main conflicting issues in product development: the need for unrestricted flow of creative pursuits and meeting the demands of the structured, planned, and timed product development commitments.

The first fully structured phase of integrated product realization is outlined in Chapter 7. Different activities such as idea enrichment, specification, synthesis, integration, development, and validation are studied. The different subphases for transforming the front-end idea kernel to an approved program by the end of the phase are identified, as well as some of the realization activities needed to complete each subphase.

In Chapter 8, the second structured phase of the integrated product realization process is outlined to include product and process, specification, planning, synthesis, integration, virtual development, and validation. The different subphases for transforming product and production process concept to prototype and production commitment are identified. Some of the virtual realization activities needed to complete each subphase are also discussed.

The last phase of integrated product realization before product launch is outlined in Chapter 9. The outline encompasses concept enrichment, specification, facilities, supply chain, physical testing, development, and validation. The different subphases for transforming prototype and production commitment to validated production facilities, pilots, and product launch are identified. Some of the realization activities needed to complete each subphase are delineated.

To enable successful implementation of integrated product realization, some of the key product realization enablers are discussed in Chapter 10, with selected enablers covering the main realization activities of analysis, integration, and management. To ensure achieving desired product and production process, characteristics, attributes, and targets the major roles for each enabler are discussed.

I.2 References

I-1. Automotive Industry Action Group. 2003. ISO/TS 16949:2002 Implementation Guide, CQI-7. Automotive Industry Action Group: Michigan.

I-2. Automotive Industry Action Group. 2009. Technical Specification ISO/TS 16949, Quality Management Systems—Particular Requirements for the Application of

ISO 9001:2008 for Automotive Production and Relevant Service Part Organizations, Third Edition.

I-3. Chrysler Corporation, Ford Motor Company, and General Motors Corporation, *Advanced Product Quality Planning (APQP) and Control Plan Reference Manual*, June 1994.

I-4. M. El-Sayed, "Engineering Design Education for Integrated Product Realization," Proceedings of the 2009 ASEE Annual Conference, ASEE Paper # AC 2009-2234, June 2009.

I-5. El-Sayed, M., "Lean Design for Integrated Product Realization," *SAE Int. J. Mater. Manuf.* 3(1):194-201, 2010, doi:10.4271/2010-01-0400.

I-6. El-Sayed, M., "Expanding Virtual Simulation in Product Realization," *SAE Int. J. Mater. Manuf.* 4(1):788-798, 2011, doi:10.4271/2011-01-0530.

Chapter 1
Evolution of Vehicle Development Process

"I invented nothing new. I simply assembled the discoveries of other men behind whom were centuries of work."

—Henry Ford

1.1 Introduction

Product development processes could be viewed as the set of steps, events, actions, or activities for transforming an idea to a product [1-1], [1-2], [1-3]. Despite some common features, product development processes may vary from one organization to another depending on the product, organizational structure, and corporate culture. These processes usually have more similarities among enterprises in the same industry. Once defined and practiced, product development processes become the habits by which corporations perform their collective activities to meet the needs of the customers and achieve their strategic goals.

To identify the sequence of events in a process, a process model is usually constructed. A process model typically represents the illustration of different phases and the activities conducted in each phase. The reasons for developing and implementing a product development process model include the following [1-4]:

- satisfying customers
- managing activities
- organizing tasks
- transferring knowledge
- tracking activities

- establishing view of operations
- identifying improvements

Due to the vital role of the development process in product and organizational success, a survey was conducted in October 2002 by the MIT Center for Innovation during a product development conference. In this survey, individuals from a variety of industries rated elements of their organizational product development. The results of this survey indicated "a great need for better implementation and execution of existing processes" [1-5]. As one of the lead industries in product development processes, the auto industry is no exception. There is always a great need for better implementation and execution of the vehicle development process. In addition, there is even more need for continuous improvement of the vehicle development process itself, which is the subject of this book.

To improve the vehicle development process, a clear understanding of the process evolution and current practices is necessary. Historically, vehicle development practices and activities have been shaped by the vehicle production mode. Therefore, to understand the evolution of the vehicle development process, vehicle production history must be studied.

In this chapter, the historical evolution of the vehicle production mode is studied. The role of the production mode in shaping the product development process is also examined. Subsequently, the evolution of the resulting process models is discussed. In addition, some of the most prominent vehicle production systems and development process models are analyzed.

1.2 Origin of Vehicle Production

The origin of vehicle production could be linked to the introduction of the wheel as the first architectural enabler, over 5500 years ago. An example of the early constructed wheels is shown in Figure 1.1 [1-6].

Figure 1.1 Example of early constructed wheels [1-6]. (arka38/shutterstock.com)

The introduction of the wheel paved the road for the development of the first chariot 300 years later. Driven by the need for better mobility, the introduction of the chariot led to the development of freight wagons, passenger coaches, and finally, the currently known vehicles.

1.3 Craft Production and Development Process

The assembly of hand-crafted pieces and components from simple ideas and concepts shaped the production mode and development process of early mobility vehicles. The chariot shown in Figure 1.2 [1-7] represents a modern copy of a hand-crafted mobility product from an early craft production process. The original gold gilded wooden chariot, exhibiting a high level of craftsmanship, was built for King Tutankhamun between 1341 and 1323 BC.

Figure 1.2 Modern copy of the hand-crafted chariot of King Tutankhamun [1-7]. (Jose Ignacio Soto/shutterstock.com)

Considering the main steps or activities followed in developing and producing early mobility products, the foundation of the vehicle development process, including the production phase, can be modeled. It is clear from the early development processes that ideas generated to address mobility needs were developed into concepts that could be hand crafted and assembled into products. Accordingly, it can be depicted that the early vehicle development process followed the three basic steps modeled in Figure 1.3.

Figure 1.3 The craft production process of early vehicles.

The process model, shown in Figure 1.3, illustrates the main phases of the development process and hand-crafted production, or simply craft production, from the start of development (SOD) to the start of production (SOP) to the craft production. Any design activity during these early processes was included in the concept development phase. Despite its simplicity, the process model of Figure 1.3 represents the product development process in its basic and most fundamental form.

1.4 Introduction of Separate Design Phase in Vehicle Development Processes

The design of a steam-powered vehicle by Nicolas Joseph Cugnot, as illustrated in Figure 1.4 [1-8], marked the introduction of self-propelled vehicles.

Figure 1.4 The design of a self-propelled steam-powered vehicle [1-8]. (Morphart Creation/shutterstock.com)

The steam-powered vehicle shown in Figure 1.5 [1-9] was constructed by the mechanic Brezin according to Cugnot's instructions for the French military in 1769.

Figure 1.5 A self-propelled steam-powered vehicle [1-9].

The development of the steam-powered vehicle did not only pave the road for today's self-propelled vehicles, but it did also document a clear standalone design phase in the vehicle development process. As the designer was different from the manufacturer, the separation of the design activities from the production activities introduced a separate design phase to the development process model, as shown in Figure 1.6. The introduction of design as a standalone phase established design as a major discipline with its own separate activities.

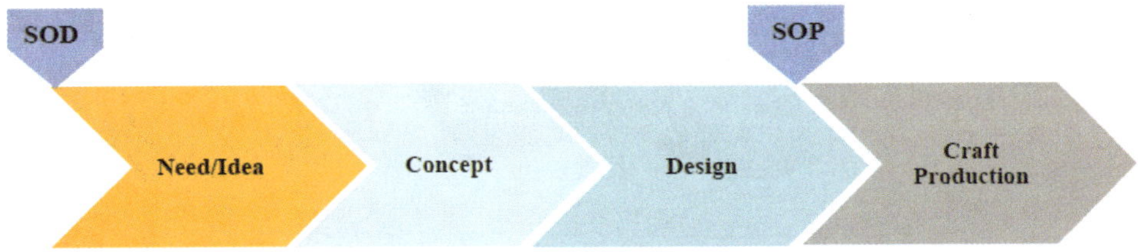

Figure 1.6 The craft production vehicle development process.

The process model, shown in Figure 1.6, illustrates the four main phases for the development process and craft production of many products, including vehicles. While concept development and design are separate phases, several variations of the model have been followed. For example, some early design activities could be started during concept development or some concept development activities could be continued through the design phase.

1.5 Mass Production and Development Process

With the introduction of the mass-produced Ford Model A in 1903, a key evolution in the vehicle development process started. About 1,700 vehicles of the first Ford Model A, exemplified in Figure 1.7 [1-10], were assembled using identical parts, in a little plant on Detroit's Mack Avenue, with 8-hp two-cylinder engines.

Figure 1.7 1903 Ford Model A [1-10]. ("From the Collections of The Henry Ford").

Without the clear construction of an assembly line, the production of the Ford Model A initiated a major transformation in auto industry by introducing the mass production concept. However, the full transformation of the production process was finally realized in 1913 with the construction of the assembly line for the mass produced Ford Model T shown in Figure 1.8 [1-11].

Figure 1.8 Ford Model T Prototype after completing 1,357 test drive miles in September 1908 [1-11]. ("From the Collections of The Henry Ford").

The Model T assembly line constructed in Highland Park, Michigan marked the beginning of a new era in vehicle development and production history. The introduction of the assembly line, as shown in Figure 1.9, marked the birth of the automated mass production and development process or simply the mass production process.

Figure 1.9 Ford Model T 1913 assembly line at Highland Park Plant [1-12]. ("From the Collections of The Henry Ford").

The continuous improvements of the assembly line by Ford, as demonstrated in Figure 1.10, reduced the assembly time of the Model T from 14 h to about 1 h and 33 min per vehicle.

Figure 1.10 Ford Model T 1926 assembly line at Highland Park Plant [1-13]. ("From the Collections of The Henry Ford").

Through automated mass production, the price of the Ford Model T was reduced from $850 to $280, making it more affordable. In fact, "affordability" of a high quality and exceptional utility, yet simple "universal vehicle," was the vision behind Ford's innovations for serving the customers. As a rule, Ford stated: "there is one rule for the industrialist and that is: make the best quality of goods possible at the lowest cost possible," which was in line with his own articulated vision: "I will build a motor car for the great multitude. It will be large enough for the family but small enough for the individual to run and care for. It will be constructed by the best materials, by the best men to be hired, after the simplest designs that modern engineering can devise. It will be so low in price that no man making a good salary will be unable to own one" [1-14].

By introducing high-quality standardized parts and assembly lines, Ford has steered the vehicle development process from craft production to mass production, as illustrated by the process model shown in Figure 1.11.

The vehicle development model outlined in Figure 1.11 illustrates the basic process phases for transforming an idea to a mass produced product. This model embodies the foundations of most mass production and processes utilized today for many products. The activities conducted during the different phases, however, may vary from one product to another and from one company to another. In general, the typical activity conducted during each phase can be outlined as follows [1-15]:

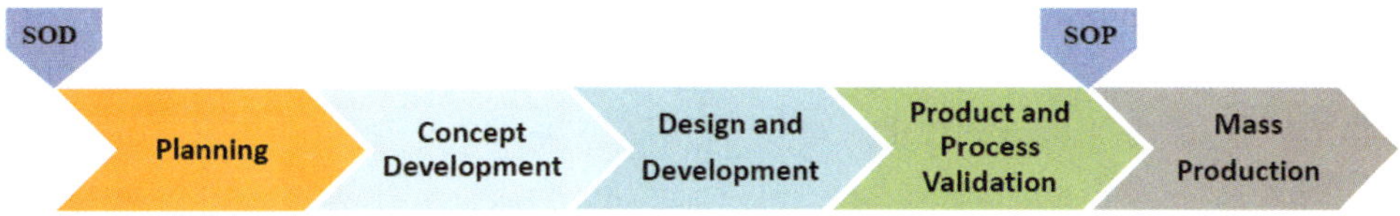

Figure 1.11 Ford mass production vehicle development process.

1. *Phase 1—Planning:* Includes need identification, idea generation, target market, business goals, key assumptions, and constraints.
2. *Phase 2—Concept Development:* Several alternative concepts for meeting the identified needs may be considered. The concept meeting most requirements and constraints is usually chosen for further product design and development.
3. *Phase 3—Design and Development:* Architectural and performance integration are conducted during this phase. Design parameters such as geometry, materials, and tolerances for all parts and components are determined by the end of this phase. Manufacturability, assembly sequences, and mass production system studies are also initiated. Some Alpha prototypes not fully produced by the intended production process are usually built during this phase for design validation.
4. *Phase 4—Product and Process Validation:* Refinement and validation of the final product and production process are completed during this phase. Several Beta prototypes and pilot vehicles are usually built with production intent tooling for final product and process validation. Production issues and workforce training are typically addressed before the SOP.
5. *Phase 5—Mass Production:* Product launch, feedback assessment and corrective actions are usually conducted at the beginning of this phase during production ramp-up.

From the basic mass production process shown in Figure 1.11, several variations have been introduced. The following are two examples of these variations:

1. Example 1.1: Separate phases for system-level design, detailed design, and production ramp-up, as shown in Figure 1.12 [1-15].

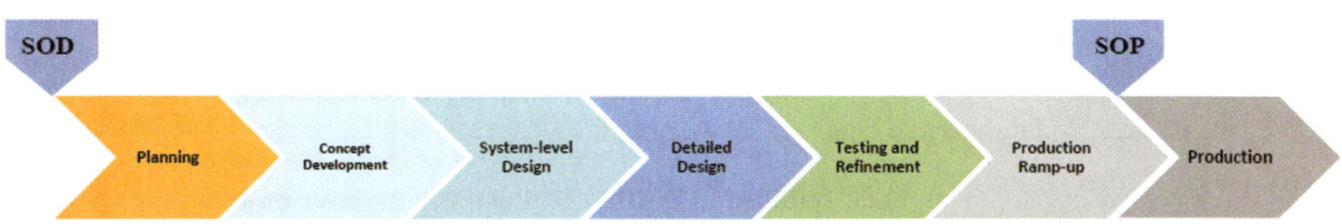

Figure 1.12 Development process with separate system-level and detailed design phases.

2. Example 1.2: Separate phases for engineering, validation, and pilot testing, as shown in Figure 1.13.

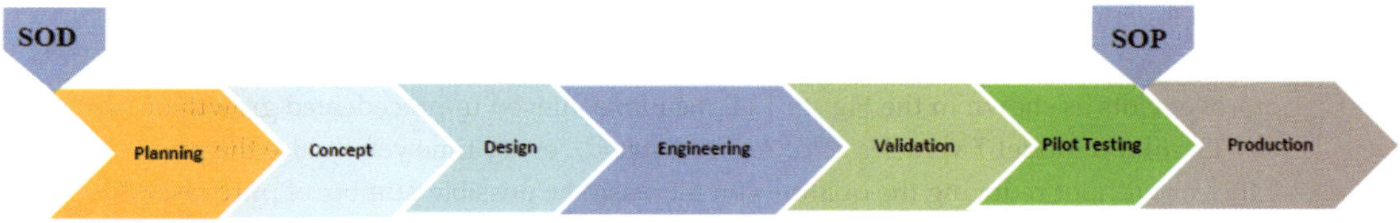

Figure 1.13 Development process with separate engineering, validation, and pilot phases.

1.5.1 Principles of Ford Production System

Automation caused the production process to be more inherently allied with the vision and principles upon which it is founded. The founding vision and principles shape the process outcomes, purpose, focus, strive, improvements, and drivers.

In the case of Ford Production System (FPS): striving for *price reduction* to serve the customers and achieve the desired outcomes of *affordability/quality*, Ford used *improved standard parts* and *increased automation* as production process improvements to push for *higher supply* of vehicles as the process key driver. Accordingly, the framework of FPS, or Ford Production "house," can be illustrated as shown in Figure 1.14.

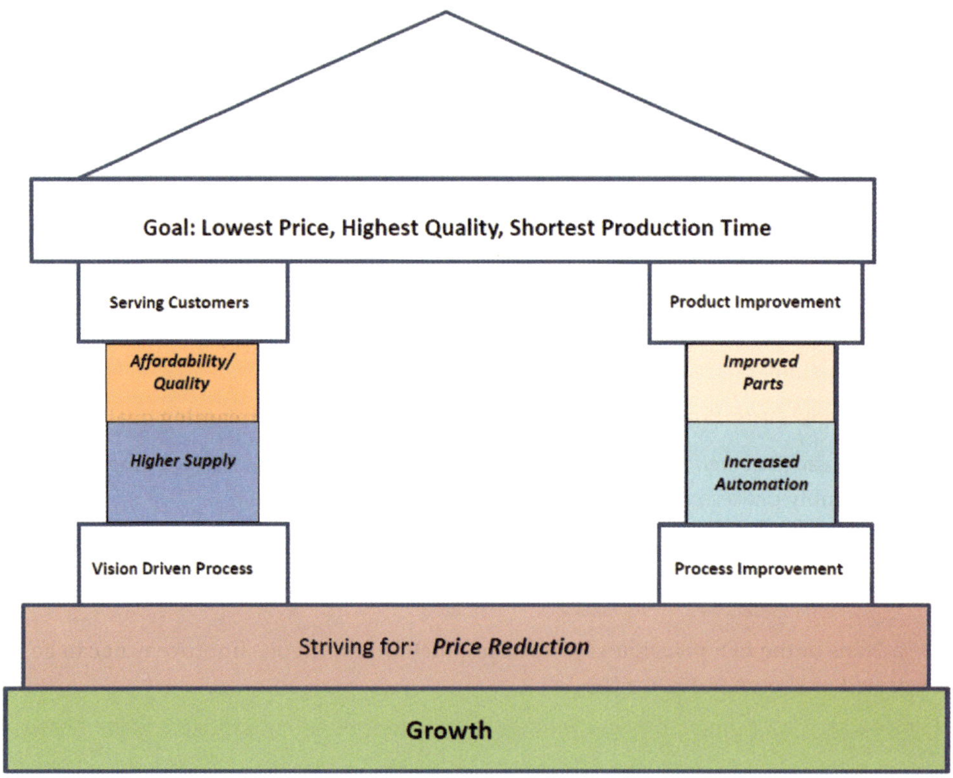

Figure 1.14 The framework of FPS.

Founded on the strive for price reduction, with one pillar representing the process outcome and driver, and the other representing the process focus for continuous improvements as shown in the Figure 1.14, the FPS achieved unprecedented growth of over 15 million Model T vehicles sold. As Ford stated, "every time you reduce the price of the car without reducing the quality, you increase the possible number of purchasers" [1-14]. By examining the FPS framework, the five principles, shown in Figure 1.15, can be specified:

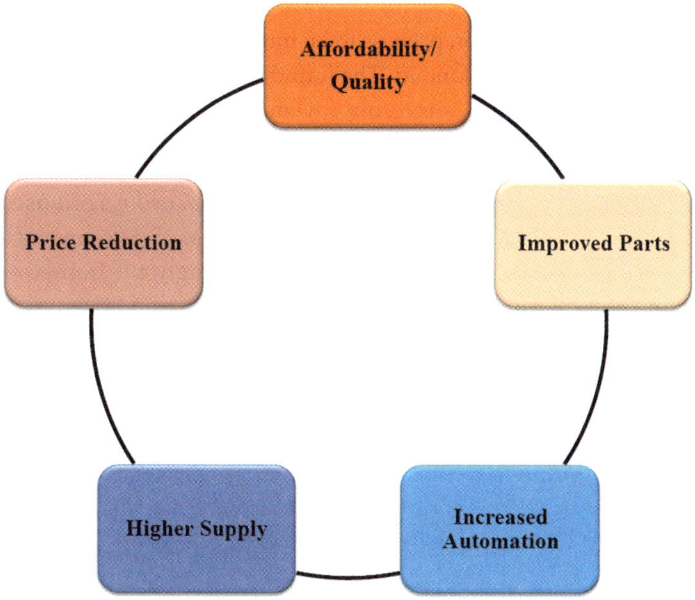

Figure 1.15 Principles of FPS.

- *Affordability/Quality:* Process desired *outcome* to realize the *purpose* of serving the customers
- *Improved Parts:* Process continuous improvement *focus* for streaming quality
- *Increased Automation:* Process continuous improvement *focus* for speeding the assembly process flow
- *Higher Supply:* Process *driver* to achieve affordability
- *Price Reduction:* Process *strive* for affordable high-quality cars

The actions of the FPS principles for mass production continuous improvement can be illustrated as shown in Figure 1.16.

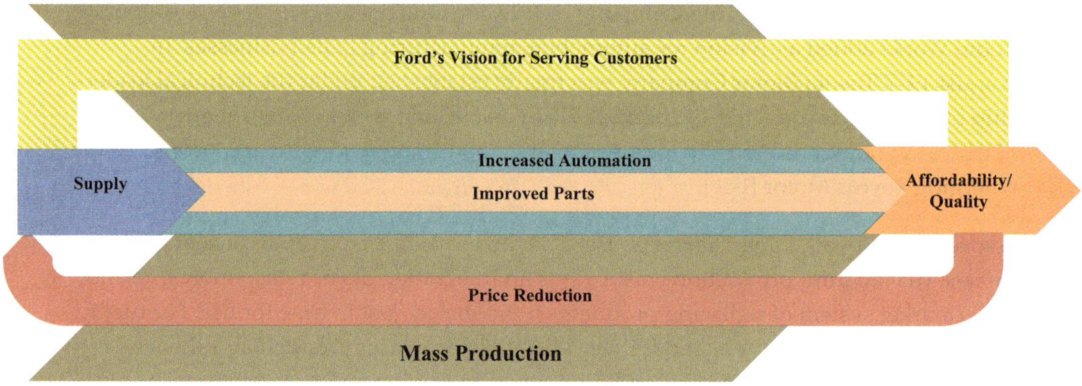

Figure 1.16 FPS continuous improvement.

The process purpose is represented by Ford's overarching vision of serving the customers. The process outcome is affordability/quality as a product characteristic while the push for higher supply constitutes the process driver. The mass production process is continuously improved through the focus on increased automation and improved parts while price reduction is the process continuous strive or aspiration.

In addition to the automated production process continuous improvement, the principles of the FPS must have been carried out through the full process, as shown in Figure 1.17, to plan, conceive, design, and develop improved vehicles. In fact, affordability and quality as product characteristics could only be achieved through implementation of the FPS principles at early planning, concept development, and design phases.

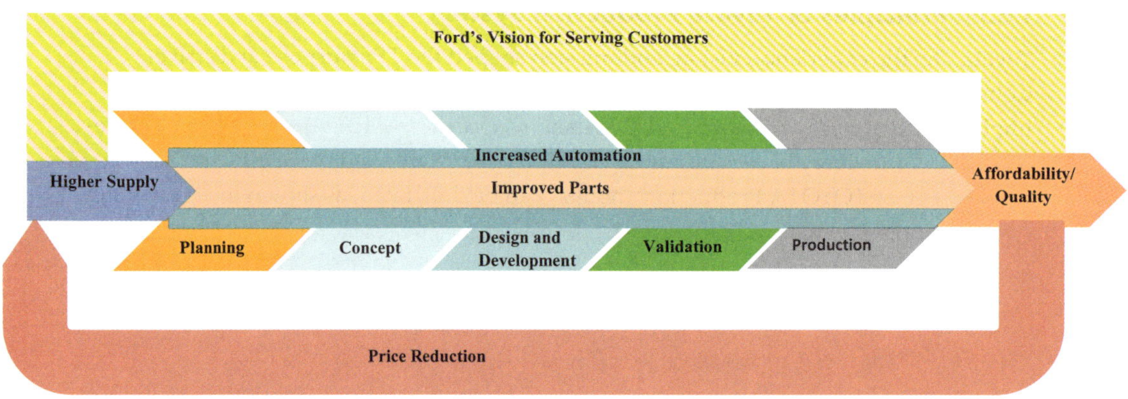

Figure 1.17 FPS and vehicle development process.

1.6 Toyota Production System and Lean Thinking

Striving to perfect his invented power looms through the JIDOKA concept of automation and quality, Sakichi Toyoda established the Toyoda Loom Works in 1906 and the Toyoda

Cotton Spinning and Weaving Company in 1918. On January 29, 1934, Toyoda Automatic Loom Works added a division for automobile manufacturing to its businesses. In August 1937, Toyota Motor Company was finally established under the technical leadership of Kiichiro Toyoda, the son of Sakichi Toyoda. Along with his management team, Kiichiro Toyoda transferred Toyoda Automatic Loom experiences and philosophy to the emerging Toyota Motor [1-16].

To overcome Toyota's financial restraints, Kiichiro Toyoda adapted the concept of just-in-time (JIT) to limit production waste. JIT is an integration of Ford's assembly line and supermarket system of replacing products on the shelves just in time for use. For further development of the JIDOKA and JIT concepts in vehicle mass production, Taiichi Ohno was reassigned from Toyoda Cotton Spinning and Weaving to Toyota Motor in 1945.

Both Toyoda Loom Works and Toyoda Cotton Spinning and Weaving had customer-driven processes with respect for customers and all involved people. As it is customary in the textile industry, products are manufactured based on customers' orders and according to customers' specifications. These products are usually made to order and delivered on time. Also, with well-established product attributes and many competing producers, as was the case in the textile industry, survival depended on continuously making things better (continuously striving for improvement), faster (reducing time via streamlined process), and cheaper (by eliminating all wastes (MUDA)). Consequently, these practices were transferred to Toyota Motor from the textile industry. These practices shaped the vision and mission of the Toyota Production System (TPS), as succinctly stated by Taiichi Ohno, the architect of the TPS [1-17]: "All we are doing is looking at the time line from the moment the customer gives us an order to the point when we collect the cash. And we are reducing that time line by removing the nonvalue-added waste."

1.6.1 Principles of Toyota Production System

Toyota's textile experience, vision, and mission shaped the TPS purpose and strive. Consequently, out of respect for people and striving for *continuous improvement*, Toyota implemented quality automation *JIDOKA* and *remove waste* (*MUDA*) as process improvements for *JIT* delivery of *customer demand*. With these established principles, the framework of TPS or TOYOTA Production "house" can be illustrated, as shown in Figure 1.18.

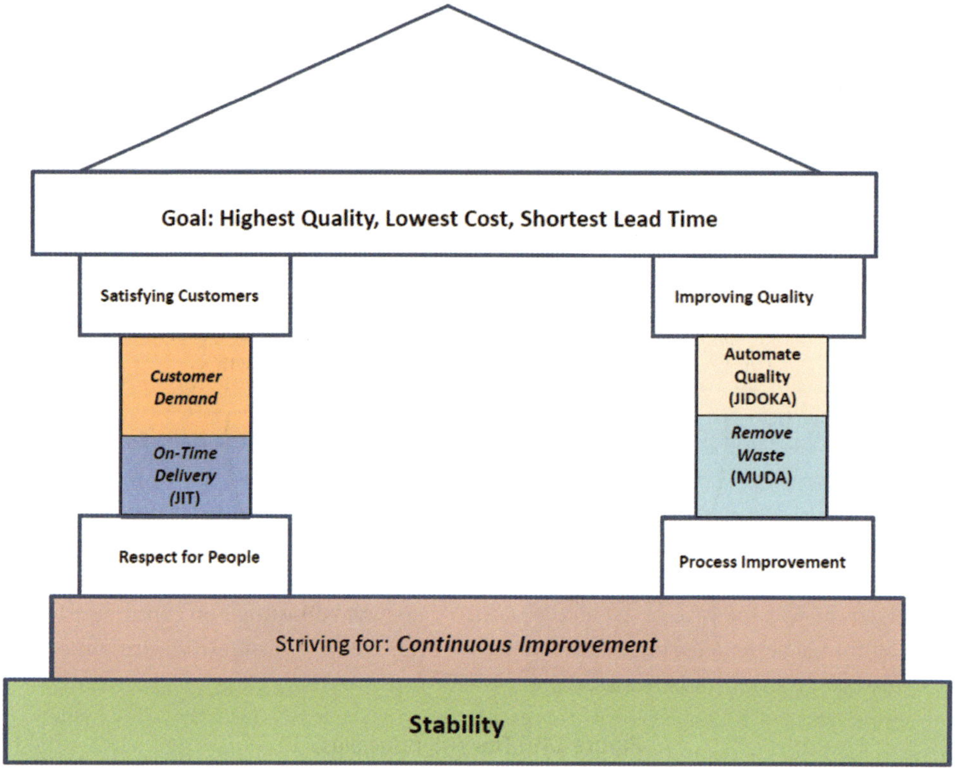

Figure 1.18 Framework of TPS.

Founded on respect for people and continuous improvement, the TPS provided stability and steady growth, elevating Toyota to the number one auto maker in the world by 2008. By examining the TPS framework, the five TPS principles, shown in Figure 1.19, can be identified as:

- *Customer Demand:* Desired *outcome* and process *purpose* out of respect for people
- *Automate Quality (JIDOKA):* A focus for process improvement by automating the quality demanded by the customers
- *Remove Waste (MUDA):* A *focus* for process improvement by removing any nonvalue-added work
- *Just-in-Time (JIT):* Process *driver* for on time delivery of customer demand
- *Continuous Improvement:* Strive for process and quality improvement

The actions of these TPS principles during the mass production phase can be illustrated as shown in Figure 1.20. The process purpose is represented by the overarching goal of serving the customers. The process outcome is customer demand, while the pull of JIT constitutes the process driver. The strive and aspiration of the TPS mass production process is continuous improvement through the focus on JIDOKA and MUDA.

As a result of Toyota's success, lean production and lean thinking were introduced [1-18], [1-19].

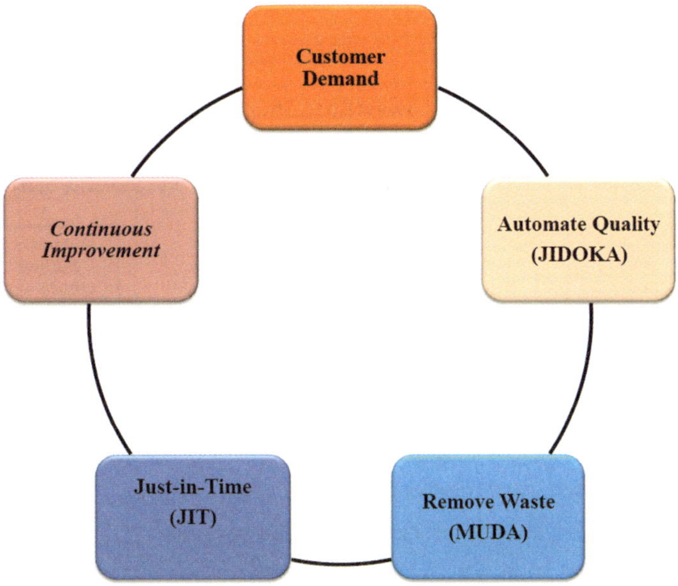

Figure 1.19 The TPS principles.

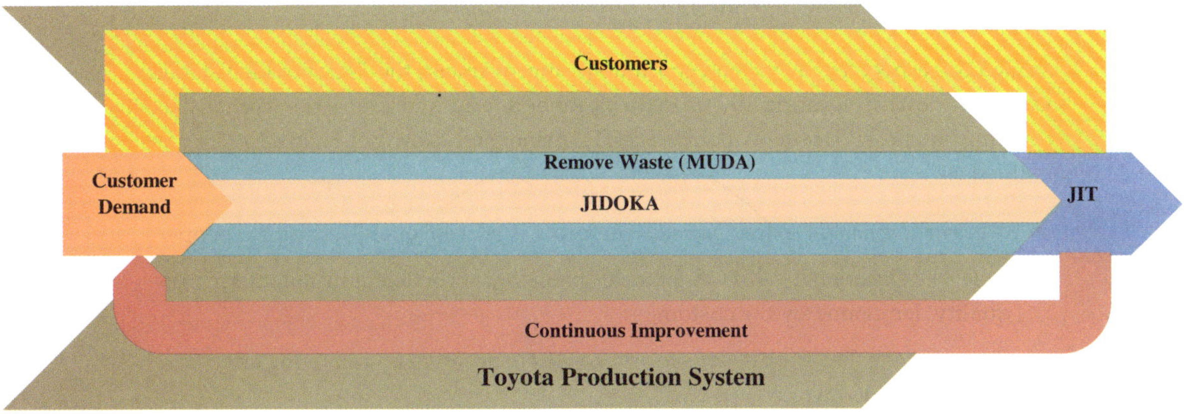

Figure 1.20 TPS production continuous improvement.

1.6.2 Principles of Lean Production System

Striving for *perfection* as a derivative of TPS, lean thinking focuses on *flow* and *value streaming* as process improvements to deliver customers' specified *value* based on customers' *pull*. With these established principles, the framework for lean thinking can be illustrated as shown in Figure 1.21.

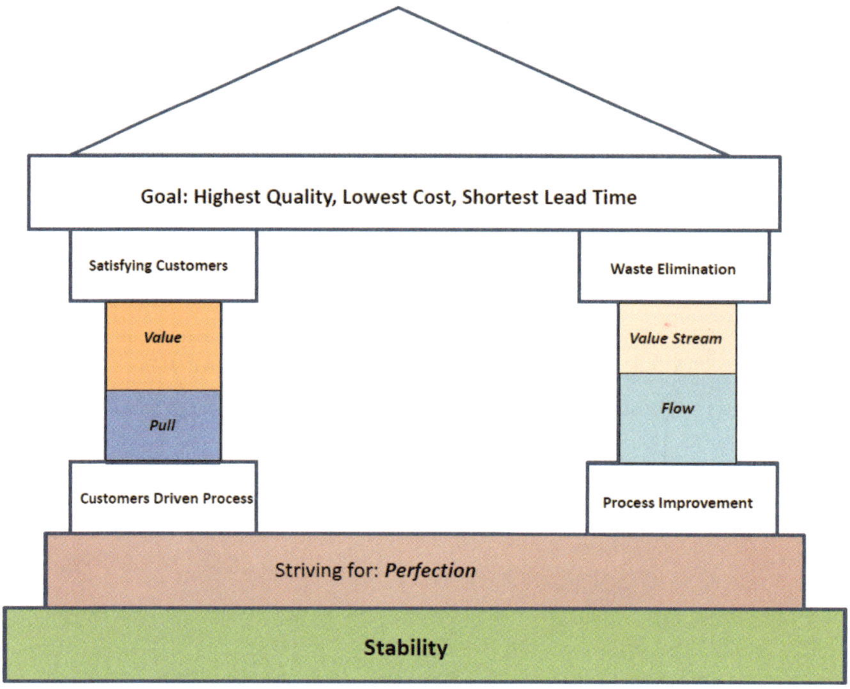

Figure 1.21 Framework of lean production system.

Based on the framework of Figure 1.19, the following five principles of lean thinking, shown in Figure 1.22, can be specified [1-18], [1-19], [1-20]:

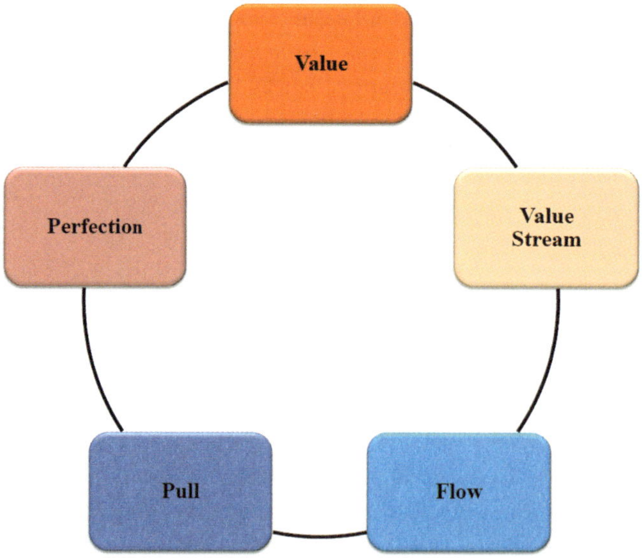

Figure 1.22 The five lean principles.

- *Value:* As specified by the customer is the desired process *outcome*, for the *purpose* of serving customers
- *Value Stream:* Streaming value along the process is a *focus* for process continuous improvement
- *Flow:* Facilitating the value flow is a *focus* for process continuous improvement
- *Pull:* Executing what is needed by the customer for JIT delivery is the process *driver*
- *Perfection:* Strive for continuous improvement of process and product

The actions of the lean thinking principles during the mass production phase can be illustrated as shown in Figure 1.23. In addition to striving for perfection, the essence of any successful lean production process is to add nothing but value. Through the focus on value, process steps or activities that do not add value are considered waste.

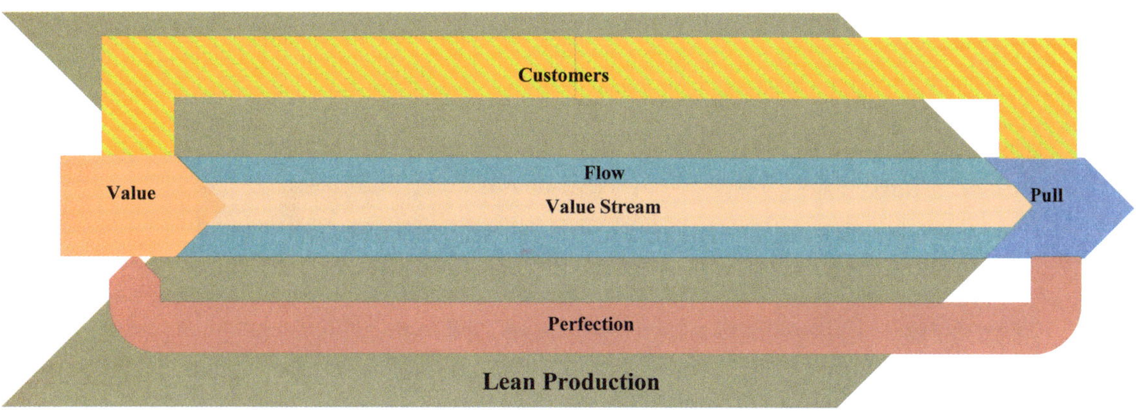

Figure 1.23 Lean production continuous improvement.

Lean thinking has provided a system-based approach for continuous improvement through its focus on value and strive for perfection [1-18], [1-19], [1-20]. Due to its flexibility and proven success in production systems, lean thinking implementation has been expanded to other industries [1-21], [1-22], [1-23], [1-24]. To fully realize value as specified by the customer, lean principles could be carried out through the full development process, as shown in Figure 1.24.

Several valuable insights pertaining to vehicle development processes and their implementations can be deducted from the FPS and the TPS. It is clear that there are some similarities between the two production systems. For example, both systems focused on process continuous improvement for achieving the desired outcomes. Also, serving the customer is the process purpose for both systems. The two systems, however, differed in how serving the customer could be achieved. FPS served the customer through an articulated vision, while TPS served the customer through seeking customer satisfaction.

FPS used product affordability of the highest possible quality as the process desired outcome, while TPS delivered what the customer specified and demanded.

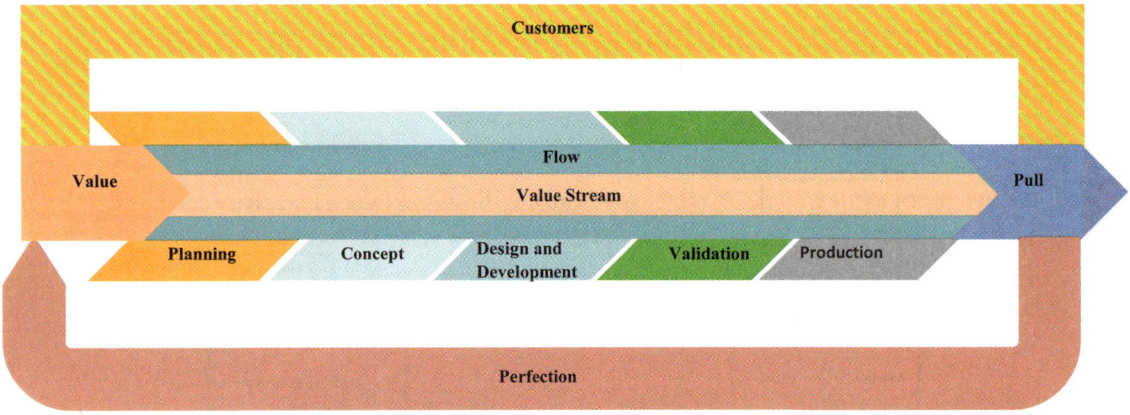

Figure 1.24 Lean production vehicles development process.

Differences between FPS and TPS on how to provide the best service to the customer can be attributed to the product, technology, and prevailing market. A clear vision for what could be achieved to serve the customers is generally suited for newly introduced products and technological innovations, as was the case with FPS. For mature products with well-established markets, providing what the customer demands is the best course of action, as was the case when Toyota started its production. Therefore, both FPS and TPS have been well suited for serving the customer within the prevailing state.

1.7 Drift/Shift in Vehicle Production and Development Process

As product development processes are the habits by which corporations perform their collective activities, a slow drift from these habits could gradually occur due to external and internal pressures. Also, due to changes in management, personnel, culture, or priorities, abrupt process shifts may occur. For instance, in addition to satisfying customers by producing innovative, low-cost, and high-quality products, auto companies must remain profitable to please stockholders and attract new investors. With the pressure of demanding quick return on investments, stockholders and investors can easily become more persistent in their demands than the customers. This imbalance in pressure may cause gradual drift or force management to sudden shift in the drive and focus of the development process. Consequently, companies may use *latest technologies* and *alluring excitements* as product improvements to push *more products*, striving for *higher sales* to achieve the desired outcome of *quick return*. Accordingly, under these pressures, the gradual drift or sudden shift in the process framework may result in an unbalanced, unsustainable, and risky production system (RPS), as illustrated in Figure 1.25.

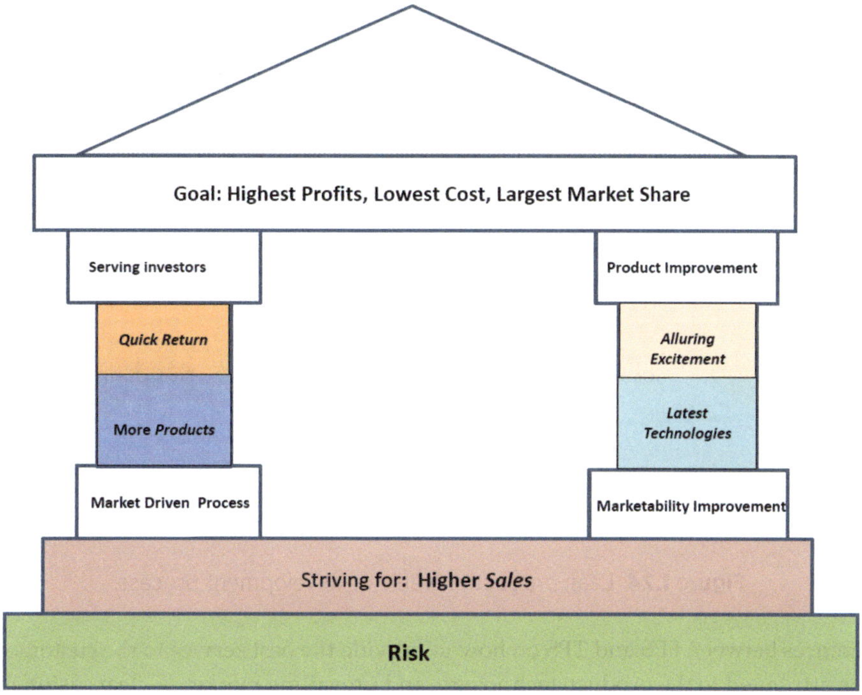

Figure 1.25 The framework of an RPS.

1.7.1 Principles of Risky Production System

By examining the framework shown in Figure 1.25, the five risky principles of a drifted or shifted production system, shown in Figure 1.26, can be identified as follows:

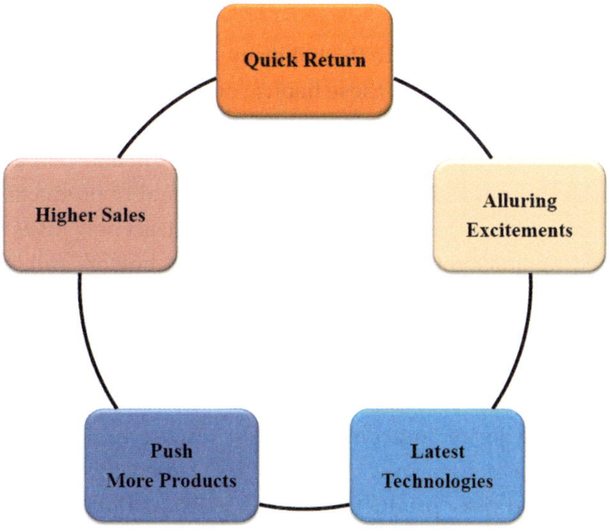

Figure 1.26 The five drifted or shifted principles of an RPS.

- *Quick Return:* Desired *outcome* for the *purpose* of appeasing stockholders and investors
- *Alluring Excitements: Focus* of improvements for attracting customers
- *Latest Technologies: Focus* of improvements to enable excitement
- *Push More Products:* Process *driver* needed to bring quick return
- *Higher Sales:* Process *strive* for achieving stockholders' and investors' demanded quick and high return

As illustrated by Figure 1.27, the effect of these risky principles is shifting or drifting of the vehicle development process.

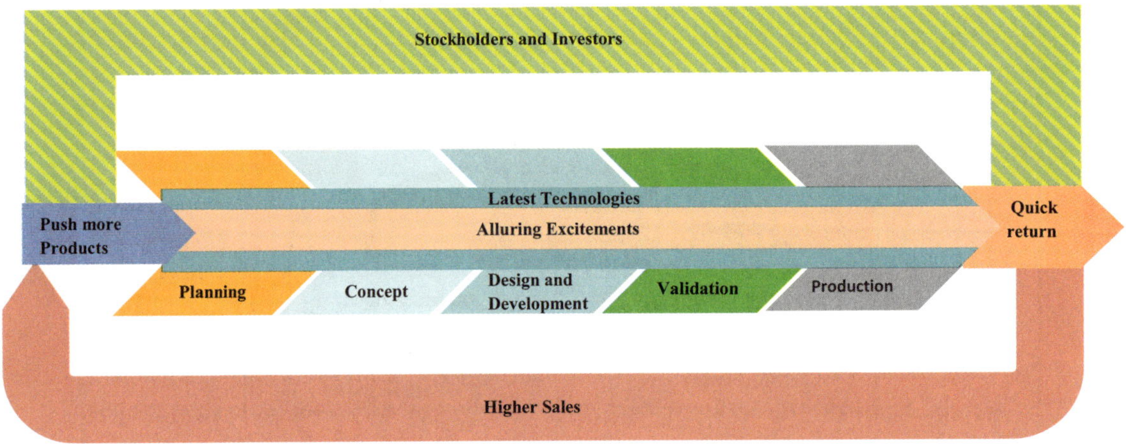

Figure 1.27 Drifted or shifted vehicle development process.

With the drift or shift in purpose from serving the customers to appeasing the stockholders and investors, quick and high return on investments become the desired process outcomes. Also, to enhance product marketability, the focus for improvement, if any, is usually drifted or shifted to product attractiveness with minimum attention to process improvements. In addition, if higher sales are not achieved due to market conditions, the pressure to cut development cost becomes higher to show return on investments. Under these pressures for cost cutting, some necessary resources for product development and validation are usually reduced or eliminated. Consequently, these organizations face higher risk due to continuous decline in product quality and customer satisfaction.

Through his experience and being aware of the resulting drift or shift on the production process, Ford stated [1-14], "The stockholders, to my way of thinking, ought to be only those who are active in the business and who will regard the company as an instrument of service rather than as a machine for making money."

1.8 Quality Planning and Management in Vehicle Production and Development Process

For quality improvements and meeting customer needs, quality planning was introduced as part of the automotive quality management system (QMS) [1-25], [1-26]. Quality planning is the activity required for determining customer needs in addition to developing the products and processes required to meet those needs [1-25]. As shown in Figure 1.28, the product realization process (PRP) is one of the core components of the QMS [1-26].

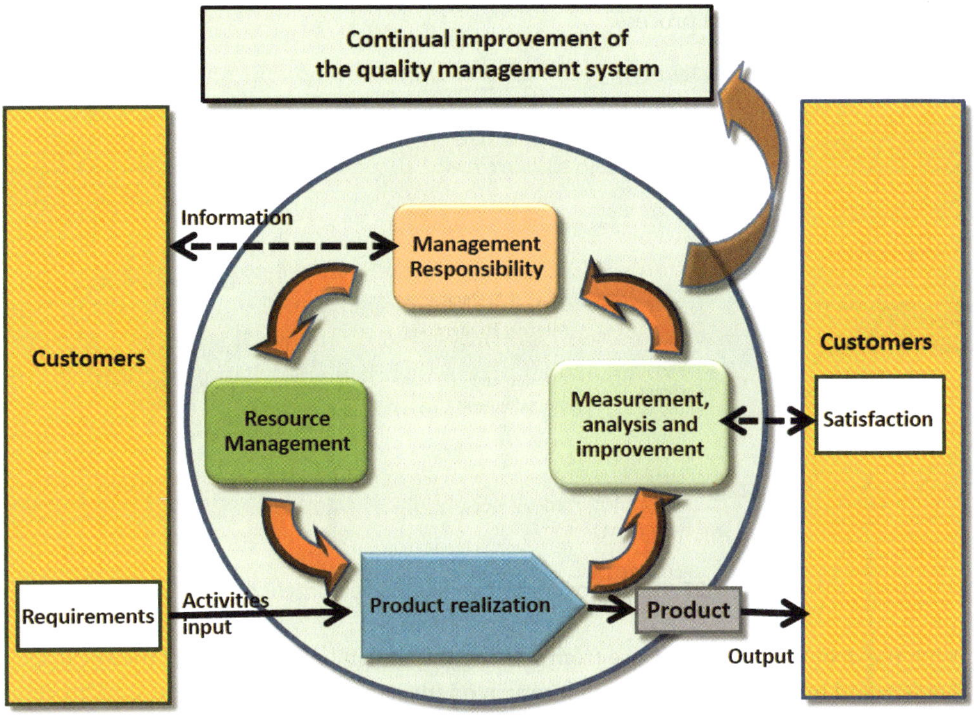

Figure 1.28 The QMS model [1-26]. ©ISO. This material is reproduced from ISO/TS 16949:2009 with permission of the American National Standards Institute (ANSI) on behalf of the International Organization for Standardization. All rights reserved. ISO/TS 16949:2009 is a withdrawn ISO document and cannot be referred to as an approved ISO standard.

The QMS model shown in Figure 1.28 illustrates the interactions between the activities of managing responsibility, resources, improvement, and product realization process. These activities are described as the sequence of actions and responsibilities that include the following areas [1-25]:

- *Management Responsibility* (*Who Owns the Process*): Represents the action of management oversight to ensure that all process steps contribute to the fulfillment of customer requirements.

- *Resource Management* (*Personnel, Skills, Equipment, Infrastructure, and Materials*): Is the provision of appropriate and sufficient resources to achieve customer satisfaction.
- *Product Realization* (*the Steps to Realize the Product*): Includes the steps for understanding customer requirements, planning, design, procurement, production, quality control, and logistics necessary to produce the intended product. Accordingly, PRP is the process by which a new product idea is conceived, investigated, taken through the design process, manufactured, marketed and serviced through obsolescence. The PRP follows the Advanced Product Quality Planning (APQP) process outlined in Figure 1.29 [1-27].
- *Measurement, Analysis, and Improvement* (*Know What You Need and What You Have*): Include validation of conformity of the process and the final product to customer requirements, and continual improvement through corrective and preventive actions.

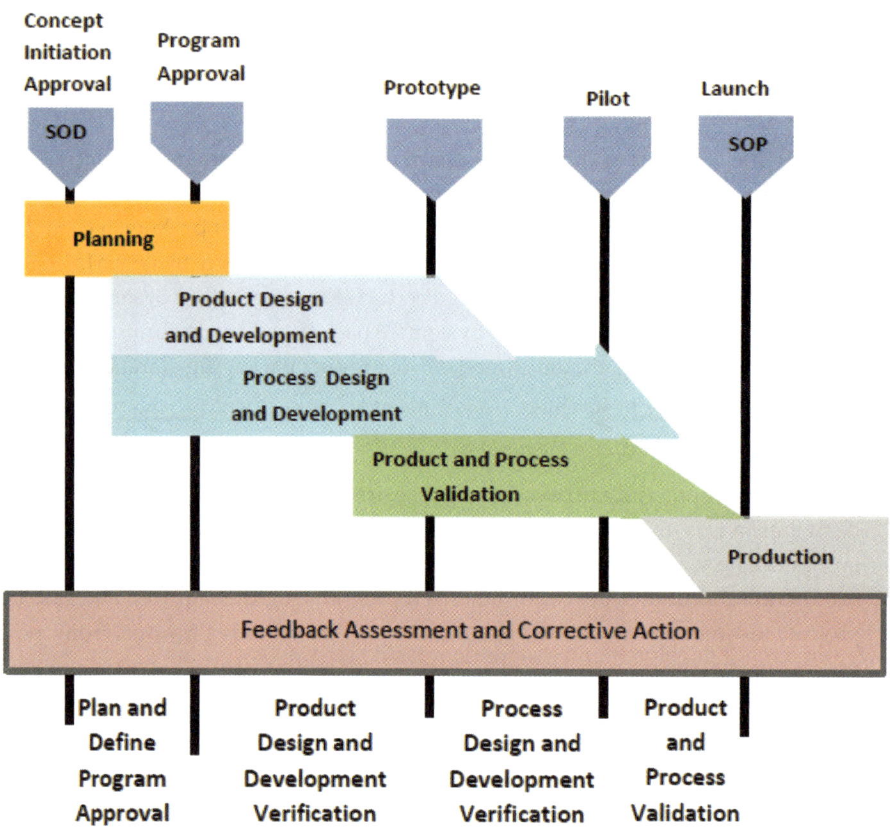

Figure 1.29 APQP product quality planning process [1-27].
(Reprinted from *Advanced Product Quality Planning (APQP) and Control Plan, Second Edition*, 2008 Manual with permission of Chrysler Corporation, Ford, and GM Supplier Quality Requirements Task Force.)

The PRP phases shown in Figure 1.29 can be summarized as follows:

- *Phase 1: Plan and Define Program:*

  This phase starts with the voice of the customer to identify customer needs, requirements, and expectations.

  The activities conducted during this phase include translating the voice of the customer into design, reliability, and quality goals in addition to the preliminary listing of special product and process characteristics. Also, the preliminary Bill of Materials and Process Flow Chart are initiated. The Product Assurance Plan transferring design goals into requirements is developed by the Product Quality Planning Team which remains in continuous communication with upper management during all phases.

- *Phase 2: Product Design and Development:*

  All product design and engineering activities which include Failure Mode and Effects Analysis (FMEA), design analysis and simulation, design verification, and design reviews are conducted during this phase. Product geometry and materials are specified by the end of this phase.

- *Phase 3: Process Design and Development:*

  The activities conducted during this phase include the design and development of the mass production process flow, capability, failure mode and effects analysis (PFMEA), floor plan, characteristics matrix, instructions, and prelaunch control plan. The Product Quality Planning Team updates the Quality System Manual of the product manufacturing site during Product/Process Quality System Review, and ensures that the Measurement Systems Analysis and the Preliminary Process Capability Study plans are developed. Also, product packaging standards and specifications are addressed.

- *Phase 4: Product and Process Validation:*

  The detailed input and output of this phase are specified by the APQP [1-27]. Activities in this phase are performed to validate the product and production process along with its control mechanisms. This phase also provides additional production conditions and requirements for identifying the required outputs. The product testing and validation are conducted on the resulting product from the process study. The outputs include objectives for manufacturing process capability, reliability, maintainability, and availability, as well as acceptance criteria.

- *Phase 5: Production Launch:*

  Feedback Assessment and Corrective Actions are core activities for this phase. Focus during this phase is directed to reducing variations, developing corrective actions, and product improvement. Finally, feedback based on customers' expectations is identified for future products [1-27]. During implementation of QMS activities and organizational interactions, the PRP customer involvement and assessment feedback can be illustrated by the process model shown in Figure 1.30 [1-25].

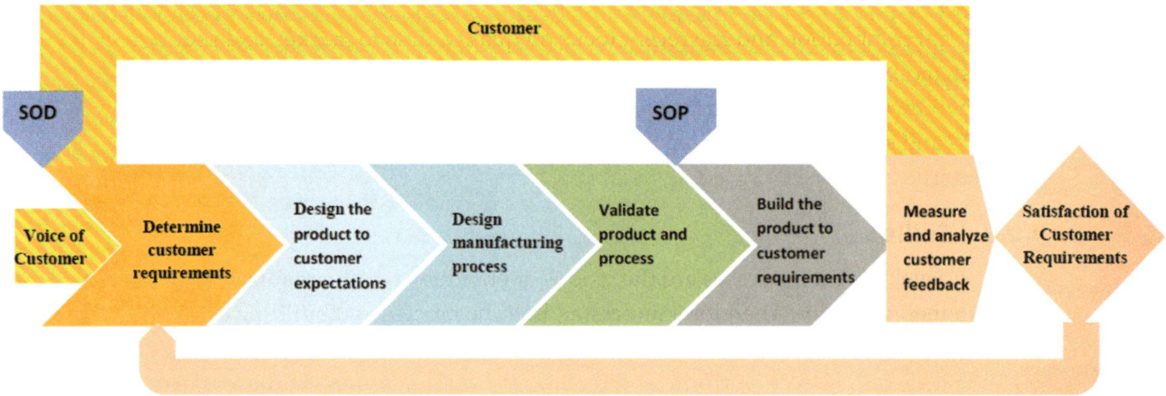

Figure 1.30 PRP and feedback assessment.

1.9 Learning Outcomes from the Evolution of Vehicle Development Process

The desired learning outcomes from studying the evolution of vehicle development process can be categorized as:

1.9.1 Knowledge

The evolution of the vehicle development process was slower than the evolution of vehicles and technologies. Until recently, the evolution in the development process has been a byproduct of the evolution in vehicles and technologies. As a result, the evolution from craft production to mass production took over three thousand years.

Early development processes started with ideas to address mobility needs. These ideas were then developed into concepts that could be hand crafted and assembled into a mobility product. For craft production systems, the speed and frequency of production are usually limited and highly controllable. In addition, needed improvements can easily be introduced to the next product with minimum or no adjustment to the process. These attributes of craft production resulted in great flexibility for product improvements, but at the same time limited the need for further process development and improvement.

With the introduction of the assembly line automation for mass production, some of the immediate control afforded by craft production was relinquished. Accordingly, automation reduced the level of immediate control over the production process. Similar to any dynamic entity, the automated mass production process outcomes became more dependent on the long-term vision and mission established by the producers instead of their mere physical presence and control. The long-term vision and mission upon which an automated process is founded define its driver, purpose, and focus. They also shape its characteristics and attributes, and ultimately determine its outcome. As demonstrated by FPS and TPS, serving the customers as process purpose and focusing on process improvements resulted in unprecedented success for both systems.

With global competition and the recent focus on quality, time to market, and production cost, more efforts have been directed to the development process in the last thirty years than the efforts made during the entire process history. The introductions of lean thinking and quality planning to the mass production process are great examples of these efforts.

1.9.2 Skills

Product development processes are the modus operandi by which organizations perform their activities. Therefore, understanding the process and building the skills needed for achieving its continuous improvement is necessary for individual and organizational success. Also, identifying, correcting, or avoiding process shifts and drifts, are additional skills needed for long-term success and competitiveness.

1.9.3 Paradigm/Attitude

Serving customers does not end with only providing the customers with what they demand but also includes providing what is the best for the customers, even if it was not demanded. For instance, if the customers admired or demanded a specific style which may cause safety issues, then sacrificing safety for the desired style does not constitute the best service to the customer. The conflict between the desired style and safety, however, presents an opportunity for creativity and innovation to achieve both. In fact, as was demonstrated by Henry Ford, striving to serve and provide what is best for the customers can be great motivating driver for creativity and innovation during the development process. As Henry Ford stated [1-14], "Some of the stockholders were seriously alarmed when our production reached one hundred cars a day. They wanted to do something to stop me from ruining the company, and when I replied to the effect that one hundred cars a day was only a trifle and that I hoped before long to make a thousand a day, they were inexpressibly shocked and I understand seriously contemplated court action. If I had followed the general opinion of my associates I should have kept the business about as it was, put our funds into a fine administration building, tried to make bargains with such competitors as seemed too active, made new designs from time to time to catch the fancy of the public, and generally have passed on into the position of a quiet, respectable citizen with a quiet, respectable business." Having the best interest of the customers at heart and being driven to achieve vehicle affordability he continued to say [1-14], "I regarded our progress merely as an invitation to do more as an indication that we had reached a place where we might begin to perform a real service."

1.10 References

1-1. Eppinger, Steven D., Daniel E. Whitney, Robert P. Smith, and David A. Gebala. "A Model-Based Method for Organizing Tasks in Product Development." Research in Engineering Design. 6(1):1–13, 1994.

1-2. Baxter, M. 1995. Product Design: A Practical Guide to Systematic Methods of New Product Development. Chapman & Hall: London.

1-3. Kennedy, M. N. 2003. Product Development for the Lean Enterprise, The Oaklea Press: Virginia.

1-4. Sanongpong, K. "Automotive Process-based New Product Development: A Review of Key Performance Metrics," *Proceedings of the World Congress on Engineering*, Vol. I, London, U.K, July 2009.

1-5. Holmes, M. F., and Campbell, R.B. Product Development Processes: Three Vectors of Improvement. Center for Innovation in Product Development, MIT: Massachusetts. 2003.

1-6. http://image.shutterstock.com/display_pic_with_logo/122248/122248,1303753631,4/stock-photo-first-stone-wheel-75933220.jpg.

1-7. http://image.shutterstock.com/display_pic_with_logo/194977/194977,1320802716,1/stock-photo-ancient-egyptian-ceremonial-chariot-isolated-with-clipping-path-modern-copy-inspired-by-those-88406539.jpg.

1-8. http://image.shutterstock.com/display_pic_with_logo/348289/401764411/stock-photo-steam-car-of-cugnot-under-louis-xv-vintage-engraved-illustration-magasin-pittoresque-401764411.jpg.

1-9. http://upload.wikimedia.org/wikipedia/commons/5/56/FardierdeCugnot20050111.jpg.

1-10. From the Collections of The Henry Ford THF 90450.

1-11. From the Collections of The Henry Ford THF 114440.

1-12. From the Collections of The Henry Ford THF 91696.

1-13. From the Collections of The Henry Ford THF 121545.

1-14. Ford, H., and Crowther, S. 1922. My Life and Work. Garden City Publishing: Garden City, New York.

1-15. Ulrich, K. T., and Eppinger, S. D. Product Design and Manufacturing. McGraw-Hill: New York. 2000.

1-16. Mass, W., and Robertson, A., "From Textiles to Automobiles: Mechanical and Organizational Innovation in the Toyoda Enterprises." Business and Economic History. 25(2):1895–1933, 1996.

1-17. Markovitz, D. Applying Lean Principles to Banish Waste and Improve Your Personal Performance. CRC Press: Florida. 2011.

1-18. Womack, J. P., Jones, D. T., and Roos, D., "The Machine That Changed the World: The Story of Lean Production." Rawson and Associates: New York. 1990.

1-19. Womack, J. P. and Jones, D. T., "Lean Thinking: Banish Waste and Create Wealth in Your Corporation." Simon & Schuster: New York. 1996.

1-20. El-Sayed, M., "Lean Design for Integrated Product Realization," *SAE Int. J. Mater. Manuf.* 3(1):194-201, 2010, doi:10.4271.2010-01-0400.

1-21. Ballard, G., and Howell, G. "What Kind of Production is Construction?" Proceedings of the Sixth Annual Conference International Group for Lean Construction, Sao Paulo, Brazil, August 1998.

1-22. Chalice, R. "Improving Healthcare Using Toyota Lean Production Methods: 46 Steps for Improvement," 2nd edition, American Society for Quality, Quality Press: Milwaukee, 2007.

1-23. Curt Hibbs, C., Jewett, S., Sullivan, M. 2009. "The Art of Lean Software Development: A Practical and Incremental Approach." O'Reilly Media: California.

1-24. Fiume, O., and Cunningham J. E. 2003. "Real Numbers: Management Accounting in a Lean Organization." Managing Times Press: Durham, North Carolina.

1-25. Automotive Industry Action Group. 2003. ISO/TS 16949:2002 Implementation Guide, CQI-7, Automotive Industry Action Group: Michigan.

1-26. ©ISO. This material is reproduced from ISO/TS 16949:2009 with permission of the American National Standards Institute (ANSI) on behalf of the International Organization for Standardization. All rights reserved. ISO/TS 16949:2009 is a withdrawn ISO document and cannot be referred to as an approved ISO standard.

1-27. Chrysler Corporation, Ford Motors Company, and General Motors Corporation, *Advanced Product Quality Planning (APQP) and Control Plan Reference Manual*, Issued June 1994.

Chapter 2
Realization Process Phases

"The locus of the human mystery is perception of this world. From it proceeds every thought, every art."

—Marilynne Robinson

2.1 Introduction

From the evolution of the vehicle development process discussed in Chapter 1, it is clear that the process has been traditionally driven and heavily dependent on the physical build for the development and validation of vehicles. The paradigm of "design–build–test" was initiated and commonly associated with vehicle development [2-1]. With the lack of virtual development tools, the physical build paradigm had dominated and constrained the vehicle development process for decades. Also, the introduction of the separate design phase resulted in the "over-the-fence" practices by having an engineering team perform the design and then pass it on to the manufacturing team. Along with other factors, the combination of "design-build-test" and "over-the-fence" practices resulted in a lengthy development cycle, costly hardware build, and unavoidable late design changes. The physical build paradigm and the segmented development process caused vehicles to be late to market, less affordable, and have many quality issues.

With the introduction and rapid development of computer-aided modeling and simulation tools in the late '70s, a gradual paradigm shift toward virtual development and validation started in the auto industry [2-2]. In fact, companies who have succeeded in reducing the design and development time have done so through the slow and steady injection of virtual modeling and simulation in the vehicle development process. While the rates of these injections have been steadily increasing, and full virtual development has been sought, most vehicle development processes still follow the physical build paradigm along with the traditionally segmented "over-the-fence" design and manufacturing practices.

Since vehicle development practices and activities are usually shaped by vehicle production mode, these practices must be altered to take advantage of rapidly evolving production systems such as additive manufacturing. Accordingly, the "design-build-test" paradigm and "over-the-fence" practices must be confronted. To shift paradigms and alter practices new paradigms, concepts, and habits must be formed. For instance, the current concepts of reality and realization and their influence on product development must be examined. As a step in this direction, this chapter aims at studying and expanding these concepts and their implications in the context of product development and realization process phases.

The term reality has been traditionally associated with the physical world. It is also well known that each individual has his or her own perception of this reality. Accordingly, with any interaction between an individual and the physical reality, there are the two identifiable domains in action. These domains are the physical domain and the individual's domain of perception. With the need to communicate and preserve the different perceptions, the third domain of collective perceptions was naturally formed. In general, this domain could be named the virtual domain. In the following, these three domains and their relation to the product realization process will be explored.

2.2 Reality Domains

When interacting with the physical world, individuals have used their perceptual senses to form their own reality. Therefore, each individual has his or her own unique representation or perception of the physical reality. To understand physical reality and collectively share their perceptions of it, individuals have utilized their communication abilities to form the virtual domain. Therefore, the definition of reality could be expanded to include other domains that may exist beyond what is physical. In the virtual domain, collective perceptions, knowledge, hypotheses, observations, and imaginations are integrated and shared. This virtual domain is nurtured and rendered in many communication forms. It may take the form of a hand drawing, sculpture, a computer graphic, a simple arithmetic, complicated mathematics, oral or written communication, etc. Therefore, the definition of reality could be expanded to include other domains that may exist beyond what is physical. Among these realities the following three domains are considered [2-3], [2-4], [2-5]:

- *Physical Domain:* Represented by the physical universe that can be recognized by the senses such as seeing, hearing, touching, smelling, and tasting. The physical domain is the domain where physical reality is being actualized.
- *Perceptual Domain:* Represented by individual perceptions or paradigms of other realities. In this domain, the individual realization is being formulated, developed,

and validated. While perceptual reality may be subjective, "perception is a form of reality" for each individual.

- *Virtual Domain:* Represented by the communicated and collectively shared perceptions of other realities. The virtual domain is the domain where collective and shared perceptions are being formulated, developed, and communicated.

In all three domains of reality, there are interacting elements or objects specified by the following:

- forms (shapes and substances)
- functions (purpose and performance)
- interactions (actions and reactions) with other object through fields of activities (interaction fields)

The transfer among the three reality domains can be considered as a sort of mapping from one domain to another. The main two elements for mapping are modeling and simulation.

2.2.1 Modeling

Modeling is the activity of creating a representation or description of an object in a domain. Depending on the reality domain, a model can be any of the following:

- a physical representation of an object (physical domain)
- a person's cognitive representation of an object (perceptual domain)
- an abstraction using oral or written communication, mathematical illustration, or computer representation (virtual domain)

2.2.2 Simulation

Simulation is the action of an object, or its model, performing in a physical, perceptual, or virtual environment. Simulation is used to determine the actual or eventual performance of an object in the actual or simulated environment. Simulations are used in all three reality domains:

- physical simulation or physical utilization of an object's physical model (physical domain)
- perceptual simulation of an object's perceptual model (perceptual domain)
- virtual simulation of an object's virtual model (virtual domain)

Mapping objects and their environment among different realities using modeling and simulation requires deducting or constructing using the following:

- analysis (for deducting or deconstructing)
- integration (for constructing or reconstructing)

Both analysis and integration use modeling and simulation to different degrees, as illustrated in Figure 2.1.

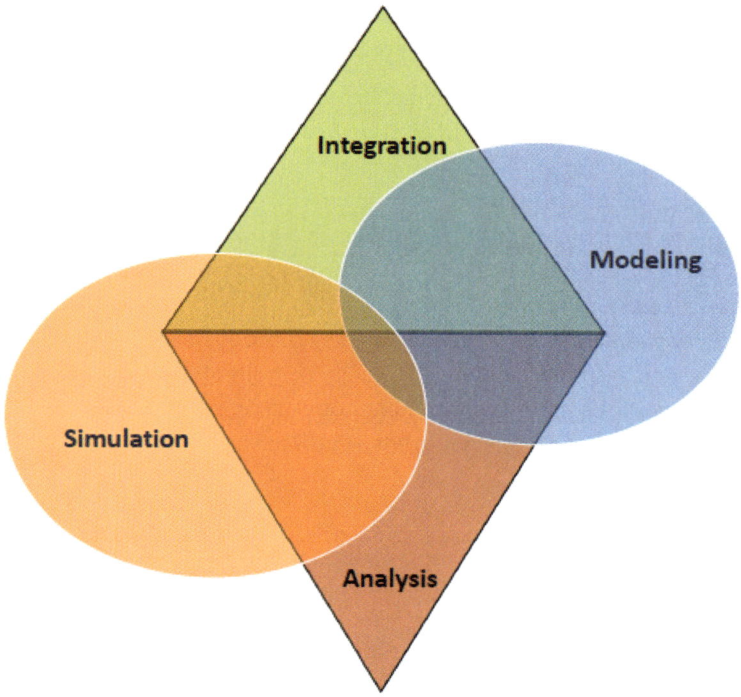

Figure 2.1 Different realization mapping processes.

As shown in Figure 2.1, while integration is performed with mostly modeling and some simulation, analysis is performed with mostly simulation and some modeling. Both analysis and simulation are mostly logical and analytical in nature, while modeling and integration are more holistic and creative in nature [2-5].

To illustrate some of the interactions among the different realities discussed so far, the following example is considered:

2.2.2.1 Example 2.1: Because passing large vehicles such as a tractor and trailer or an agricultural vehicle is more difficult than passing passenger vehicles, the U.K. Highway Code [2-6] states that a driver should perform the following:

- Drop back to increase the ability of seeing ahead and making sure to be seen in the mirrors of the large vehicle driver
- Make sure that there is enough room to complete the passing before starting
- Not to follow other vehicles in passing large vehicles because they may abort and pull back

As shown in Figure 2.2, the passing driver must map the whole reality of the large vehicle size, speed, and surrounding traffic from the physical to the perceptual domain. Through the perceptual modeling and simulation abilities (interaction between the physical and perceptual domains), the driver must first analyze the situation by creating a perceptual model through integrating all perceptual inputs and then perform a perceptual simulation to make a decision. The driver must also integrate all the guidelines stated in addition to any other present situations, such as speed limits, before attempting the passing process.

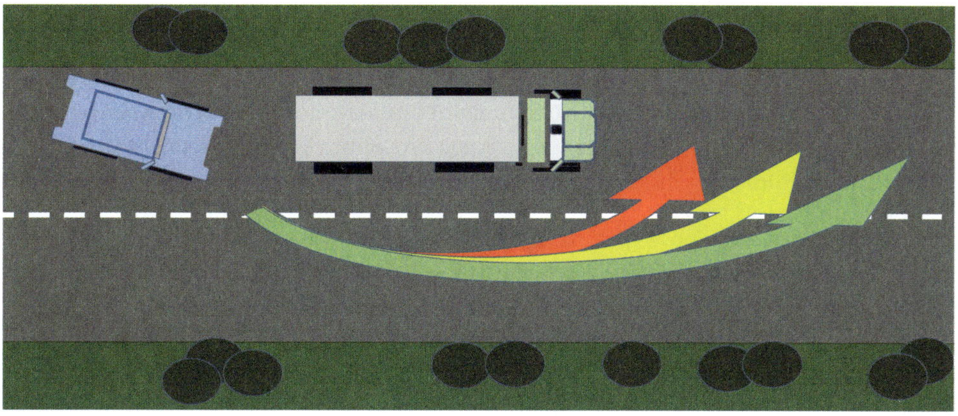

Figure 2.2 Perceptual modeling and simulation in passing a large vehicle.

During the passing process, the driver's visual images are virtual representations of the physical reality. Also, any input, such as a warning by a passenger accompanying the driver in the passing vehicle, based on his or her perceptual realization, can be considered as an additional virtual input to the driver. Integrating this virtual input in the modeling and simulation processes represents an interaction between the virtual and perceptual domains, which may alter the final realization of the passing process in the physical domain.

This example demonstrates how the full realization of the passing process in the physical domain was completed through the modeling and simulation in the driver's perceptual domain before the execution. The level and speed of mappings, simulations, analyses, and integrations among the reality domains in this simple example illustrate some of the undetected skills providing the abilities to deal with a multitude of similar daily occurrences. Carrying this example to different contexts of higher performances, such as race car driving or fast paced sports, reveals the levels of talent and skills needed for mapping and simulations between the different domains.

2.3 Realization

By examining the three reality domains and how a reality is constructed, in one domain or across all three domains, the following theory of realization or realization hypothesis can be made:

To realize is to attain a state of comprehensible reality in a domain.

Accordingly,

Realization is the state of attaining a comprehensible reality in a domain.

Consequently,

Realization process is the process for attaining a comprehensible reality in a domain.

In general, the coherence and validity of the constructed reality depend on the rigorous validation to which the reality had been subjected. With this definition of realization, there are many ways to classify the different types of realizations. For instance, realization can be classified as holistic versus analytical [2-7], [2-8], [2-9], [2-10], [2-11]. It can also be classified as receptive versus imaginative [2-12], [2-13], [2-14], [2-15], [2-16], [2-17], [2-18], [2-19], [2-20], [2-21]. As a function of time, realization can also be classified as spontaneous versus gradual [2-22], [2-3], [2-24]. The expression "I just realized," for example, is usually stated when a certain input completes the gradual attainment of comprehensible perceptual reality or when new or sudden perceptual realization is attained (paradigm shift). Given these classifications, learning in most educational settings can be categorized as analytical, receptive, and gradual perceptual realization. Arts, on the other hand, can be classified as the virtual or physical holistic, imaginative, and in some cases, spontaneous perceptual realizations. As for product realization, a combination of all types of realizations spanning all three reality domains could be involved. The product realization process could encompass holistic and analytical, imaginative and receptive, spontaneous and gradual perceptual, virtual, and physical realizations.

2.3.1 Realization Objectives

Active interaction with any of the three realities discussed is usually driven by one of the following three objectives:

- to understand it (know it)
- to utilize it (use it)
- to alter it (improve it)

These three objectives are interconnected and overlapping because utilizing or altering a reality requires some understanding of it. Also, utilizing or altering a reality brings a better understanding of it. As shown in Figure 2.3, the three realization objectives match the cognitive domain development phases of progression toward higher maturity levels [2-25]. These levels are:

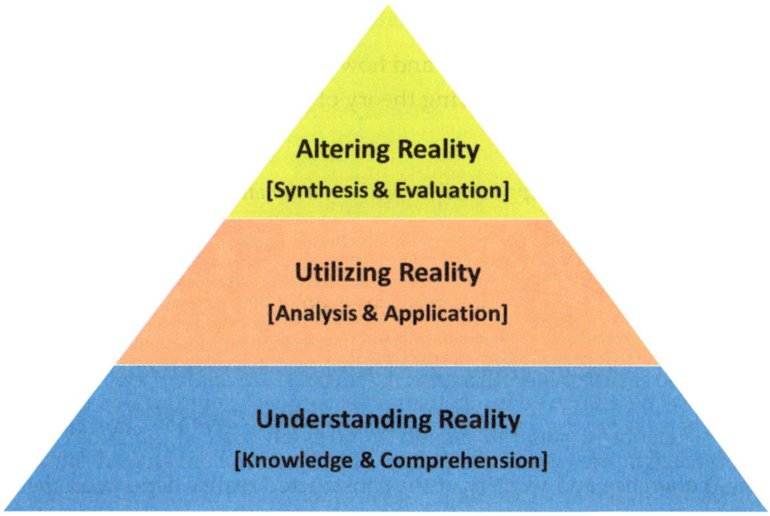

Figure 2.3 Progression of realization objectives and cognitive domain maturity [2-25].

1. knowledge and comprehension
2. analysis and application
3. synthesis and evaluation

2.3.2 Realization Activities

For achieving the three realization objectives, the following three examples of distinct but interconnected realization activities can be identified as illustrated in Figure 2.4:

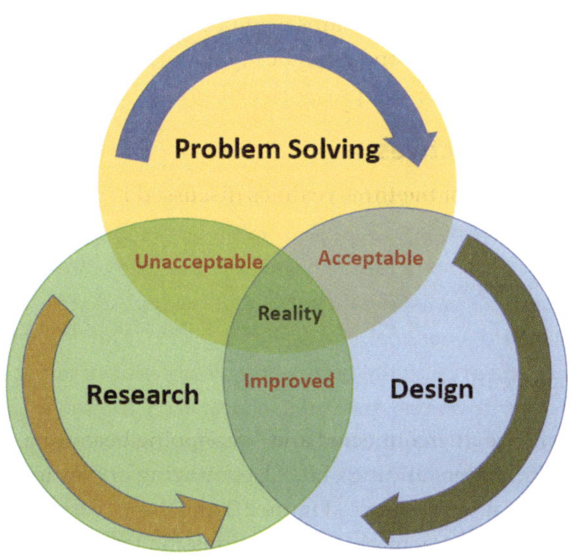

Figure 2.4 Beginning and end states for different activities.

1. *Research (Education):* To understand reality
2. *Problem Solving (Restoration):* To utilize reality
3. *Design (Creation):* To improve or alter reality

These activities can take place in any reality or across all three realities. Each of the three activities may be started due to one of the following states and could be ended by reaching another state, as shown in Figure 2.4. These states are:

1. unacceptable reality
2. acceptable reality
3. improved or altered reality

The arrows shown in Figure 2.4 indicate the direction from the beginning state to the desired or goal state for each of the three activities. As shown in Figure 2.4, the three activities are overlapping. Keeping in mind that incidental discovery could occur during any one of the three activities, the following is a brief summary of each [2-3].

2.3.2.1 Research (Education): Research, as well as education of accumulated research findings, is a perceptual activity aimed at altering (enhancing) perceptual realization of a specific subject or object. While a research activity may be conducted in the virtual or physical domain, the goal state is always an enhanced realization at the perceptual domain. As shown in Figure 2.4, research is usually triggered due to an unacceptable state of understanding of a specific reality related to a subject, an object, a function, or interaction in one of the three reality domains. Research ends when the desired goal state of perceptual realization or understanding of this reality is reached.

The research findings or new discoveries are eventually communicated and added as new knowledge in the virtual domain. Through education, the perceptual realizations of the learners are then constructed with the collectively accumulated knowledge. Therefore, research and learning are the foundations for building the perceptual domain realization.

2.3.2.2 Problem Solving (Restoration): In the context of realization, problem solving aims at restoring or transforming a reality from the unacceptable state to the acceptable (normally expected) state as shown in Figure 2.4. For example, in the physical domain, a problem exists if a car does not stop when the brake is applied. The expected reality or state is that the car should stop when the brake is applied. The problem solving, in this case, is restoring or transforming the nonworking brake reality to the expected state of a working brake reality. In general, problem solving is conducted using the perceptual domain to solve problems at any of the other domains. The solution, however, is always realized at the domain where the problem started.

2.3.2.3 Design (Creation): Design or creation is a creative and analytical activity conducted by the perceptual domain aiming at altering (enhancing) the virtual or the physical domains by introducing new or modified desired objects. Design or creation is usually triggered due to an acceptable or unacceptable state of reality that needs to be or could be altered. The end goal state of the design or creation is an improved reality. While the goal for research is to alter reality at the perceptual domain, the goal for design or creation is usually to alter reality at the physical or virtual domain to a better state. For new product creation, design activity represents the virtual realization during the product development process through which a new product is created in the physical domain.

It is important to recognize that while the starting domain can be different for each activity, all three activities are conducted by the perceptual domain. It is also important to understand when each of these activities begins and ends. For example, many product research or design activities turn into problem solving, and some problem solving may turn into research or require new design [2-3]. To illustrate the realization objectives and the difference between realization activities, the following example is considered.

2.3.2.4 Example 2.2: To improve the performance of an engine, some research activities are needed. This research would start due to an unacceptable state of understanding the engine performance and continue until an improved state of understanding is reached. Utilizing the perceptual domain of the researchers and conducting the research in the physical and/or virtual domains through testing physically and/or virtually simulated models of the engine, could produce an improved state of realization in the researchers' perceptual domain. If the research results are documented and communicated or published, an improved state of realization in the virtual domain would result as well.

During the research activities, if an engine part or an object in the physical domain was not working as expected, a problem would be identified by the perceptual domain. To develop a solution may require the utilization of the virtual or physical domains in addition to the perceptual domain until the problem is solved.

Finally, the improved perceptual understanding gained during the research and/or problem solving activities could be utilized to design and create a new engine with improved performance in the physical domain.

2.3.3 Realization Phases

Realization process follows the maturity continuum of three interconnected but identifiable phases. As shown in Figure 2.5, these phases can be identified as inception, conception, and maturation. Similar to many natural systems, the subphases for each realization phase are repeats of the identifiable formation of inception, conception, and maturation in a fractal-like pattern.

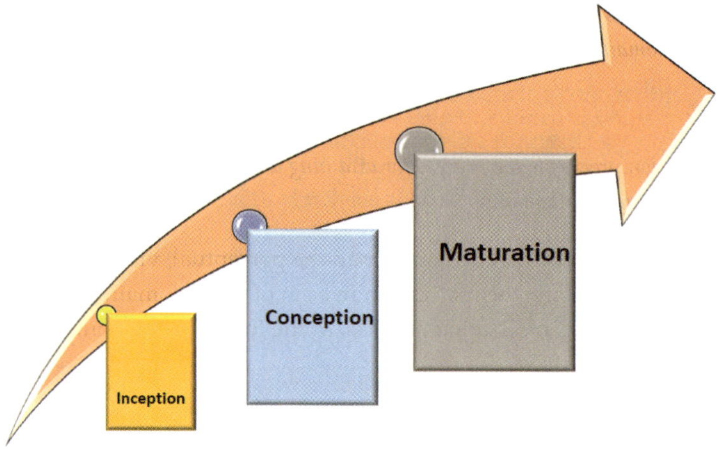

Figure 2.5 Phases of product realization process.

2.3.3.1 Inception: Initialization and specification. It starts with recognizing the features and ends with a clear and complete specification of these features (recognizing and specifying features).

2.3.3.2 Conception: Synthesis and integration. It starts with assembling the individual features and ends with integrating these features (creating and joining specified features).

2.3.3.3 Maturation: Development and validation. It starts with finalizing the integrated features and ends with verifying or validating the realized outcome (finalizing and confirming realized outcome).

Inception signifies the beginning or the start of the realization process. Conception embodies the synthesis and fusion of incepted elements into a concept. Maturation is the phase of maturing a concept to a fully developed and validated realization. While the steps or phases are similar, the shape of the outcome is domain dependent. For example, inception in the perceptual domain takes the form of the start of an idea or ideation, and conception is the formation of an abstract concept or a mental image. In the virtual or physical domain, a concept is the rendering of the synthesized realization into a virtual or physical model.

2.4 Product Realization

The term product realization has been traditionally used to describe the physical outcome of a production or manufacturing process. Accordingly, product realization meant bringing a physical product to reality. In ISO 9001, the term was expanded to include all product development activities, as stated: "Product realization is the term used to describe the work that the organization goes through to develop, manufacture, and deliver the finished goods or services" [2-26]. Following the realization hypothesis discussed earlier, further specification to the concept of product realization can be made:

Product Realization is the state of attaining a comprehensible product reality in a domain or across multiple domains.

Accordingly,

Product Realization process is the process for attaining a comprehensible product reality in a domain or across multiple domains.

Consequently, product realization could span the perceptual, virtual, and physical realization domains. Therefore, the realization in each of these domains could constitute a phase in the full product realization process. In the following section, these product realization phases are discussed.

2.4.1 Product Realization Phases

Product realization spanning the perceptual, virtual, and physical domains can be viewed as gradual, imaginative, partially holistic, and partially analytical realizations. As a realization process, product development follows the inception, conception, and maturation phases modeled in Figure 2.6 and described below:

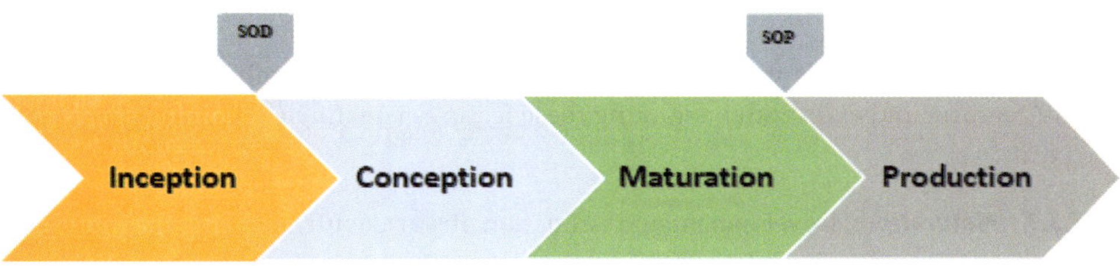

Figure 2.6 Natural product realization process phases.

- Product inception is the start of defining the desired or required product attributes (defines and specifies the desired product attributes).
- Product conception is the synthesis and fusion of attributes to realize a desired product concept (integrates and forms the product concept).
- Product maturation is the development and validation of the realized concept for production and delivery (matures and confirms the achievement of the desired outcomes).

Consequently, in product realization, inception signifies the collective perceptual domain fully realized, developed, and validated product idea. Conception embodies the synthesis and fusion of the collective perceptual domain and the fully realized idea into a new virtually and/or a physically validated design or product concept. Maturation designates the development of the concept into a new physically validated product ready for mass production. In the following, the product realization phases and subphases spanning the three realization domains will be further detailed.

2.4.1.1 Perceptual Phase (Product Inception Phase): Perceptual realization is mainly holistic, driven by customers' desired attributes. The collective perceptual product model, created during this phase, addresses the architectural characteristics and attributes such as styling, harmony, and packaging. It also addresses the product performance characteristics and attributes such as safety, durability, and utility. As a realization phase, the perceptual phase follows the three interconnected but identifiable subphases, shown in Figure 2.7 and described below:

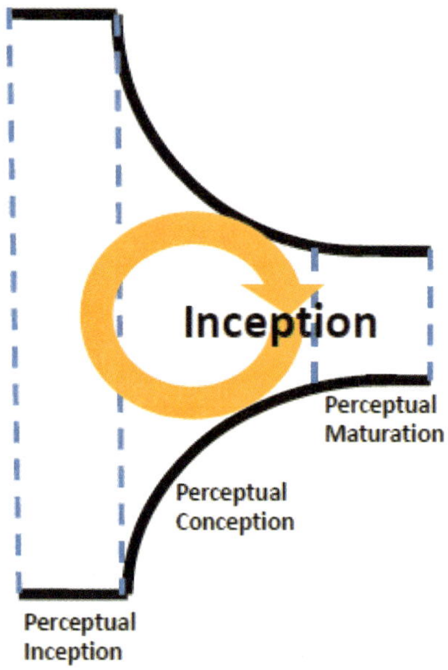

Figure 2.7 Inception realization phase.

- Perceptual inception is the start of collecting the customers' desired product attributes (gather attributes).
- Perceptual conception is the fusion of product attributes to synthesize the product idea (put attributes together).
- Perceptual maturation is the development and validation of the synthesized product idea into a validated product plan (confirm what incepted could be virtually and physically realized).

Readiness for the next phase can be validated by a virtually simulated model and/or physical mockup.

2.4.1.2 Virtual Phase (Product Conception Phase): Virtual realization is a mix of holistic and analytical realizations, driven by established product targets. The product virtual or physical concept, created during this phase, addresses product and process targets. As a realization phase, the virtual phase also follows the three interconnected but identifiable subphases, shown in Figure 2.8 and described below:

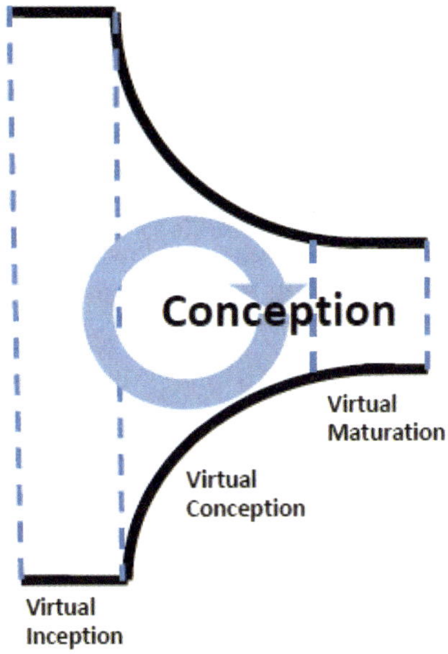

Figure 2.8 Conception realization phase.

- Virtual inception is the start of identifying and selecting desired or required product and production subsystems to be virtually modeled (gather needed subsystems).
- Virtual conception is the synthesis and fusion of the virtual subsystems to create product and production concept (put subsystems together).
- Virtual maturation is the development and validation of the virtually realized product and production concept (make sure the realized concept is working).

Readiness for the next phase could be validated virtually or by a physically simulated and physically built concept model (Alpha build—for design validation or physical proof of concept).

2.4.1.3 Physical Phase (Product Maturation Phase): Physical realization is mainly driven by established specifications of the product, subsystems, and components. The physically developed product, subsystems, and components must meet all specified physical tests for validation. As a realization phase, the physical phase follows the three interconnected but identifiable subphases, shown in Figure 2.9, and described below:

- Physical inception is the start of identifying and selecting the desired or required physical product and production subsystems and components (gather physical product and production subsystems and components).
- Physical conception is the synthesis of physical product and production subsystems for final production (put physical product and production subsystems together).

- Physical maturation is the development and validation of the physically realized product and the production system (make sure that physically realized product and production systems are working).

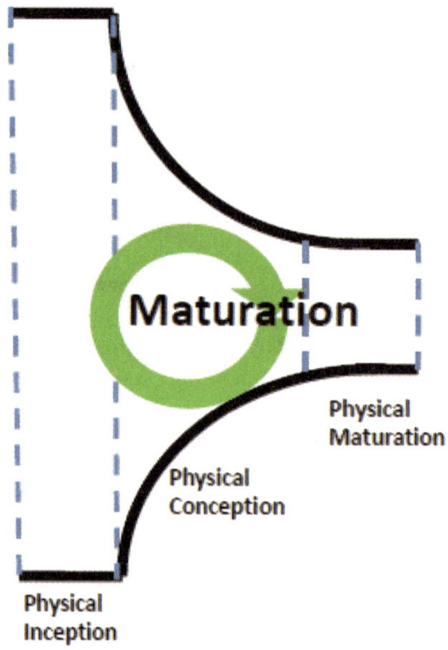

Figure 2.9 Maturation realization phase.

Readiness for the next phase can be validated by a physically simulated production model (pilot build—for product validation and physical proof of production intent).

Pilot vehicles are finally built with production tooling for physical product and process validation. Production workforce training and production issues are mostly addressed by the end of this phase before the start of production.

2.4.2 Sequential and Parallel Implementation of Product Realization Phases

The three product realization phases discussed can be implemented sequentially with concurrency between the overlapping subphases. In a sequential implementation, as shown in Figure 2.10, the perceptual maturation and the virtual inception subphases can be performed concurrently. Also, the virtual maturation and the physical inception subphases can be performed concurrently.

From the process flow, it can be seen that besides simplicity and similarities between the phases, each phase represents a standalone module assuring the process modularity. Also, some of the activities in each subphase can be performed concurrently [2-27].

For sequential implementation, customers' feedback for quality continuous improvement and product revisions can be implemented as shown in Figure 2.11 [2-27].

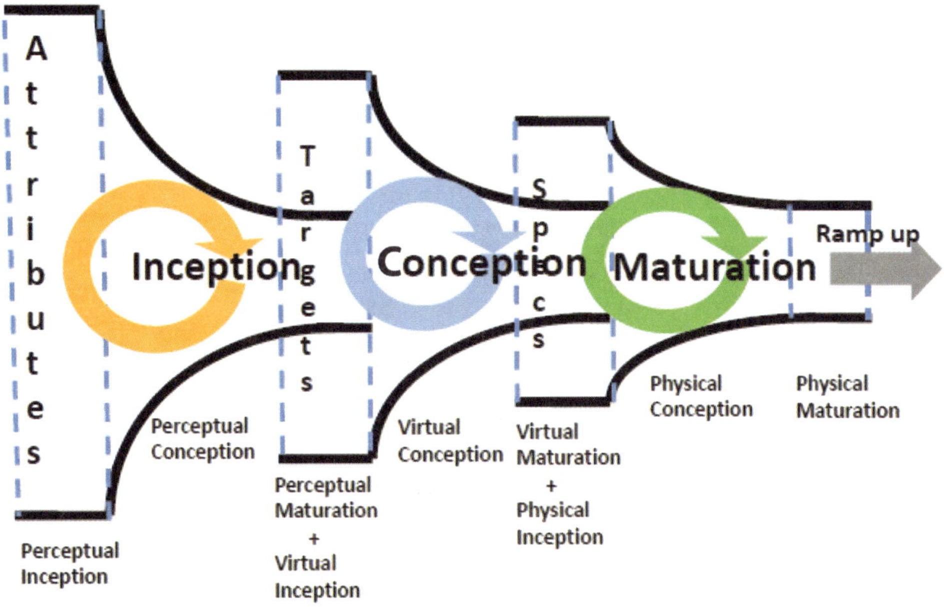

Figure 2.10 Natural product realization process phases.

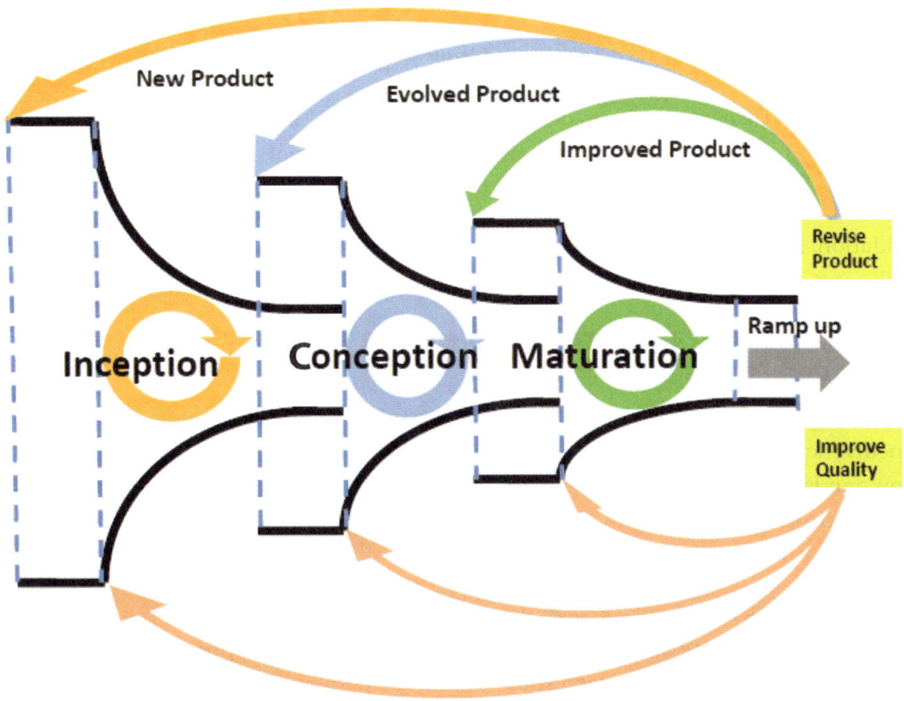

Figure 2.11 Sequential realization process with feedback and product revisions.

In Figure 2.11, the top arrows represent product revisions, while the bottom arrows represent customer inputs for quality improvement. Also, by creating an idea portfolio at the end of the inception phase and a concept portfolio at the end of the conception phase, the three product realization phases can become totally modular and could be performed in parallel, as shown in Figure 2.12.

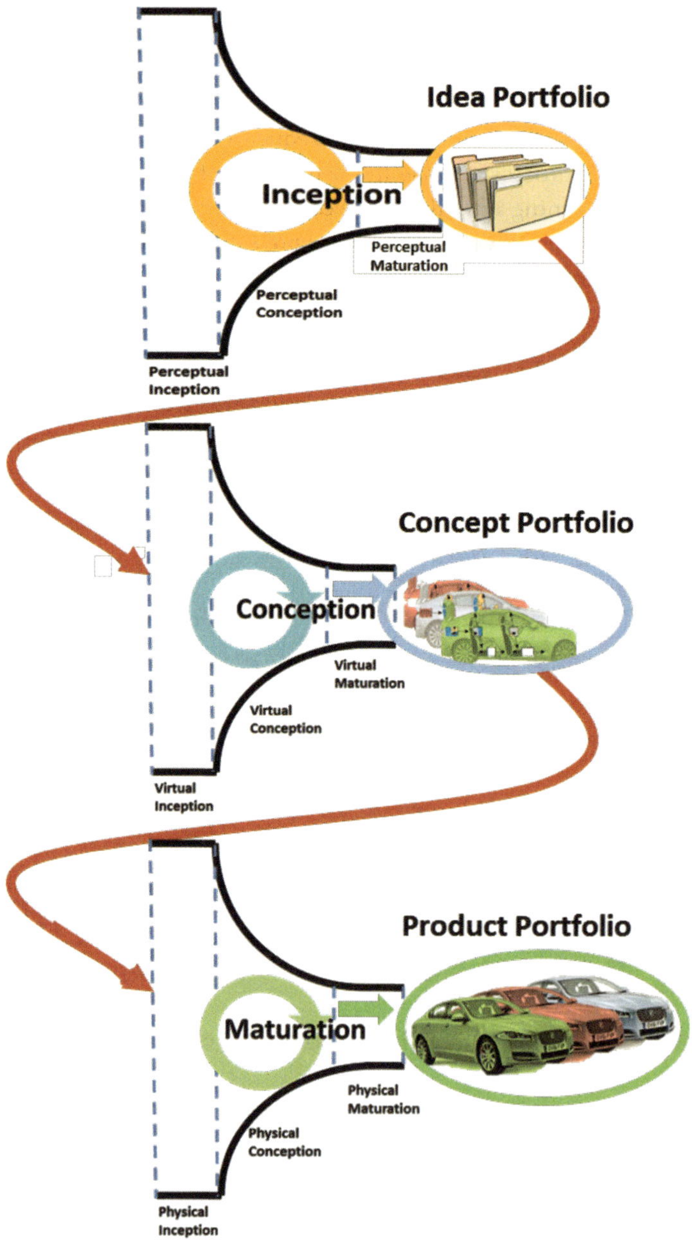

Figure 2.12 Parallel realization process with portfolios.

In parallel realization, incepted ideas are added to the idea portfolio, and validated concepts are added to the concept portfolio. The conception phase starts with a selected idea from the idea portfolio, and the maturation phase starts with a selected concept from the concept portfolio. During the parallel implementation of product realization, only validated product ideas can be selected from the idea portfolio for conception. Other ideas can be recycled for further improvements. Similarly, only validated product concepts can be selected from the concept portfolio for maturation and production. Concepts that are not selected can be recycled for further improvements.

Following the different phases of the product realization process, from inception to production, the full process can be viewed as a sequence of mappings from one reality domain to another. The process starts with the mapping of the perceptual realization during product planning activities to the collective virtual realization domain for carrying out the design activities. Finally, the physical realization or production is achieved through the mapping from conception or design, as the shared virtual domain, to the mass production physical domain through the maturation activities.

As discussed before, modeling and integration are more holistic, imaginative, and creative activities, while both analysis and simulation are more receptive, logical, and analytical realization activities. A product realization process typically starts with imaginative holistic realization at the inception phase, and ends as receptive analytical realization at the maturation phase, as shown in Figure 2.13.

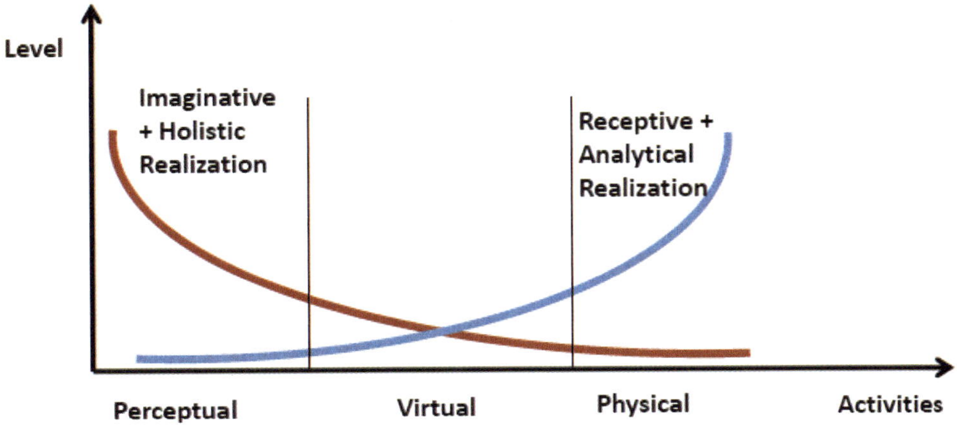

Figure 2.13 Different levels of realization types during product development phases.

The process technical execution at the perceptual phase is mainly modeling and integration with limited analysis and simulation. The perceptual inception subphase is mainly focused on holistic creative activity for the beginning of inception. As the process moves into the synthesis mode of the perceptual conception subphase, holistic creative modeling is led by styling. Subsequently or concurrently, the performance integration activities led by engineering follow. The activities during this subphase may require some perceptual analysis and simulation. The development and validation of the inception maturation subphase usually require additional analysis and integration involving some virtual and/or physical modeling and simulation.

The process technical execution at the virtual phase starts with a decelerating higher level of modeling and integration and ends with an accelerating higher level of analysis and simulation. At the start of the virtual inception, which may be concurrently performed with the inception maturation, modeling and integration are the main activities with limited but increasing analysis and simulations. As the virtual modeling and integration of the product and production process are continued with a decreasing rate during the virtual conception or synthesis subphase, the virtual analysis and simulation are increased. The increase in virtual analysis and simulation activities continues during the virtual maturation phase until all product and process targets are achieved. The development and validation of the virtual maturation subphase usually require proof of concept using physical product simulations involving craft production physical builds.

The process technical execution during the physical phase is mainly analysis and simulation with limited modeling and integration. During the physical inception, which may be concurrently performed with the virtual maturation, virtual and physical analyses and simulations are the main activities with some modeling and integration of physical builds. Physical and virtual analysis and simulations are usually the main activities during the synthesis or physical conception phase for both product and production facilities. Modeling and integration activities are usually reduced as the development of the product and production facilities are nearing completion. The increase in physical and virtual analysis and simulation activities continues during the physical maturation phase until all validation and production targets are achieved.

Considering that the physical simulations are the most costly and time-consuming activities, during the realization process, it becomes clear that validating the perceptual and virtual realizations can reduce physical builds and result in significant cost and time savings. Completing perceptual activities to the extent that the realized product idea is well developed and validated can not only reduce time and cost but also foster creativity, improve quality, and ultimately increase customer satisfaction.

Similarly, further time and cost savings can be achieved by completing the virtual activities to the extent that realized product concept is well developed and validated. Considering that physical simulations during virtual realization are usually the most costly and time-consuming activities (due to the need for special parts, tools, and craftsmanship), using virtual modeling and simulations can produce significant cost and time savings. Using correlated virtual tools during physical realization could also reduce unnecessary physical builds. However, totally replacing physical and pilot builds necessary for validating key characteristics and attributes such as safety, reliability, and durability with uncorrelated virtual simulations can have a drastic adverse effect on the product performance and quality.

To demonstrate the potential time-to-market reduction that can be achieved using virtual modeling and simulations, the following case study is considered.

2.4.2.1 Case Study 2.1: Reduction of Lead Time at Volvo: To reduce product development lead time from concept to start of production, "Volvo is turning to virtual testing and CAE to replace time-consuming physical tests" [2-28]. Volvo's goal is to reduce the lead time from 42 months to 30 months by 2015 and to 20 months by 2020 [2-28].

Using virtual modeling and simulation can add many advantages to the realization process including the following [2-5]:

1. better understanding of the design space, constraints, and limitations
2. ability to lead the realization process and examine several alternatives
3. ability to optimize the product and production process
4. faster turnaround time and shorter product realization cycle
5. cost effectiveness and higher return on investment

With the rapid development of computational technologies, many physical simulations could be replaced through continuous correlations between virtual and physical modeling and simulations. For example, a simulation process was developed to reduce the time needed for virtual frontal crash simulation of a full vehicle [2-29].

While expanding virtual modeling and simulation activities may be very desirable, addressing product validation needs is also necessary. As discussed earlier, the realization process can be viewed as a sequence of mappings from the perceptual domain to the virtual domain and finally to the physical domain. The physical domain, however, contains multiple physical realities. For example, there is a difference between the product developer simulated physical reality and the physical realities in the field of operation. In fact, the physical reality of each customer is as unique as the customer's own perception. Therefore, without correlations and clear understanding of the product and customer usage variations, many field failures and quality issues could result.

Physical build activities can be improved using advanced technologies and production tools. For example, the emerging and rapidly developing additive manufacturing or 3-D printing can drastically simplify the product realization process. In fact, product development using additive manufacturing highlights the importance of the realization concepts discussed in this chapter. Through physically validated virtual conception, physical realization could be achieved with 3-D printing of the final product. Consequently, the development process from inception to conception to maturation and finally production is direct mapping between the three reality domains.

As customers' expectations and perceptual valuations are the main factors in the assessment of quality, customers' involvement at different stages of the development process is imperative. The timing for these customers' involvements is critical for having meaningful assessments when product and process improvements are possible.

By examining the quality management system and product realization process discussed in Chapter 1, it is clear that the customers' participation is at the beginning and end of the new product development cycle. While the customers' participation at the beginning is very effective and necessary for establishing product characteristics and attributes, the participation at the end of the process is usually too late for taking any corrective actions during the process execution. As the realization process phases are a sequence of mappings among perceptual, virtual, and physical domains, including customers' participation and review at the end of each realization phase, as shown in Figure 2.14, could provide an opportunity for customers' assessment and validation feedback before proceeding to the next phase. The added customers' assessment and validation would minimize product and market uncertainty in addition to avoiding quality issues and costly corrections later in the process [2-30].

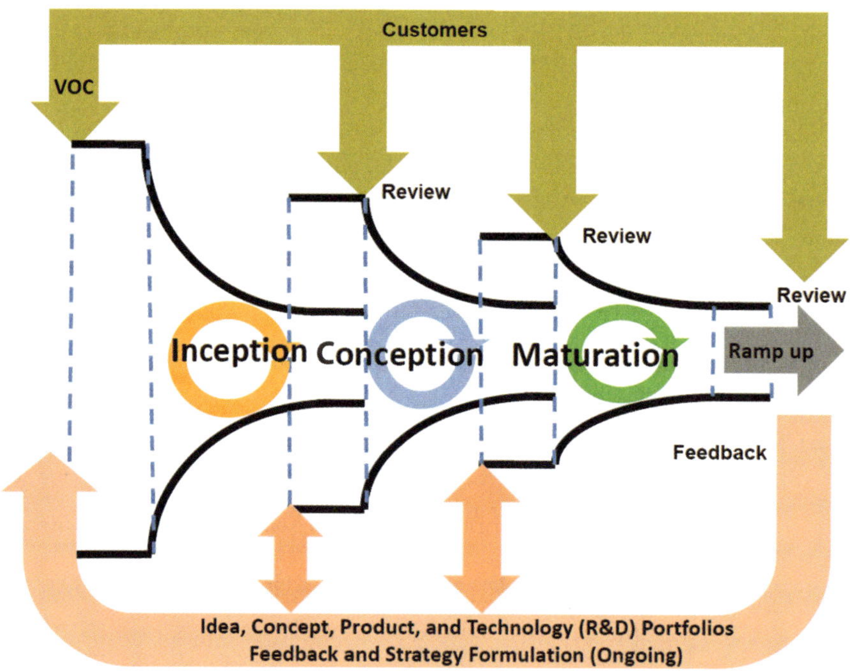

Figure 2.14 Natural product realization process with assessment feedback.

2.5 Learning Outcomes from Realization Process Phases

The desired learning outcomes from studying reality domains and realization process phases can be categorized as follows.

2.5.1 Knowledge

To communicate and share perceptions of the physical reality, the virtual reality domain was introduced. Consequently, there are three types of realities: the perceptual reality which is unique for each individual, the virtual reality which is shared between communicating individuals, and the physical reality which is universal. Considering that realization is the process of constructing a comprehensible reality in a domain or across multiple domains, the acts of realization have some similarities among the three reality domains. These similarities can be useful in understanding and constructing different realizations.

As realization activities, research or education, problem solving or restoration, and design or creation are performed to understand, utilize, and alter reality, respectively. All the three activities use analysis and integration through modeling and simulation as the basic elements for realization and mapping among realities. While analysis and simulation are mainly analytical and logical activities, modeling and integration are more holistic and creative activities.

In product realization, inception signifies the collective perceptual domain fully realized, developed, and validated product idea. Conception embodies the fusion of the collective perceptual domain fully realized idea into a virtually and/or a physically validated product concept or design. Maturation designates the development of the concept into a validated product ready for mass production. During a product development process, from an idea to physical production, conceptualization is the mapping of the perceptual realization to the collective virtual realization domain. Therefore, design is the virtual realization of a product. Physical realization or production is the mapping from design at the shared virtual domain to the physical domain of manufacturing for mass production.

For having meaningful assessments when product and process improvements are possible, customers' valuation and review can be added at the end of each realization phase. This could provide an opportunity for customers' feedback and validation of the product idea or concept before proceeding to the next phase. Consequently, quality issues, costly corrections later in the process, and uncertainty could be minimized.

2.5.2 Skills

For learning or building a coherent perceptual realization, the skills needed are perceptual analysis and integration using perceptual modeling and simulation. Similarly, for product realization, the skills needed are product integration and analysis using product modeling and simulation. These skills are also necessary for efficient mapping among the different reality domains and successful product realization. Therefore, in addition to the perceptual skills, specific virtual and/or physical analysis and integration skills using virtual and/or physical modeling and simulation tools may be needed for different activities.

2.5.3 Paradigm/Attitude

For product realization, the paradigm shift or attitude adjustment is realizing that physical reality is not the only realization domain. Perceptual realization is the source and beginning of any physically realized product. For complex products needing the knowledge and skills of different perceptual realities, the virtual realization or design is the intermediate domain for constructing the needed collective coherent realization. This collective realization must be validated before a mature product and mass production process can be realized.

2.6 References

2-1. Schäfer, H., and Sörensen, D., 2007. "The Automotive Development Process: A Real Options Analysis," Springer: New York.

2-2. Martin, A., and Monteiro de Carvalho, M. "Applying Virtual Simulation in Automotive New Product Development Process." Product: Management & Development. 4(2):79–85, 2006.

2-3. M. El-Sayed. "Engineering Design Education for Integrated Product Realization," *Proceedings of the 2009 ASEE Annual Conference*, ASEE Paper # AC 2009-2234, June 2009.

2-4. El-Sayed, M., "Lean Design for Integrated Product Realization," *SAE Int. J. Mater. Manuf.* 3(1):194-201, 2010, doi:10.4271/2010-01-0400.

2-5. El-Sayed, M., "Expanding Virtual Simulation in Product Realization," *SAE Int. J. Mater. Manuf.* 4(1):788-798, 2011, doi:10.4271/2011-01-0530.

2-6. https://www.gov.uk/guidance/the-highway-code/using-the-road-159-to-203.

2-7. Kemler, Deborah. G., 1983. "Holistic and analytic modes in perceptual and cognitive development." *Perception, cognition, and development: Interactional analyses*, pp. 77–102.

2-8. Foard, Christopher F., and Deborah G. Kemler. "Holistic and Analytic Modes of Processing: The Multiple Determinants of Perceptual Analysis." *Journal of Experimental Psychology: General.* 113(1):94–111, 1984.

2-9. Choi, Incheol, Minkyung Koo, and Jong An Choi. "Individual Differences in Analytic Versus Holistic Thinking." Personality and Social Psychology Bulletin. 33(5):691–705, 2007.

2-10. Shepp, Bryan E., and Soledad Ballesteros. 2013. "Object Perception: Structure and Process." Psychology Press: Hove, U.K.

2-11. Sattler, Rolf. 2012. "Biophilosophy: Analytic and Holistic Perspectives." Springer Science & Business Media: Berlin.

2-12. Schilder, Paul. "Mind: Perception and Thought in Their Constructive Aspects." Columbia University Press: New York. 1942.

2-13. Podgorny, Peter, and Roger N. Shepard. "Functional Representations Common to Visual Perception and Imagination." *Journal of Experimental Psychology: Human Perception and Performance.* 4(1):21–35, 1978.

2-14. Gordon, Rosemary. "Imagination as Mediator Between Inner and Outer Reality." *The Arts in Psychotherapy*. 12(1): 11–15, 1985.

2-15. Pylyshyn, Zenon. "The Role of Location Indexes in Spatial Perception: A Sketch of the FINST Spatial-Index Model." Cognition. 32(1): 65–97, 1989.

2-16. Young, J. Michael. "Kant's view of imagination." Kant-Studien 79(1–4):140–164, 1988.

2-17. Brann, Eva TH. 1993. "The World of the Imagination: Sum and Substance." Rowman & Littlefield Publishers: Lanham, Maryland.

2-18. Joy, Annamma, and John F. Sherry. "Speaking of art as embodied imagination: A multisensory approach to understanding aesthetic experience." *Journal of Consumer Research*. 30(2): 259–282, 2003.

2-19. Allison, John. 2003. "A Way of Seeing: Perception, Imagination, and Poetry." SteinerBooks: Massachusetts.

2-20. Gallagher, Michael Paul. "Theology and Imagination: From Theory to Practice." Christian Higher Education. 5(1): 83–96, 2006.

2-21. Seeley, William P. "Naturalizing Aesthetics: Art and the Cognitive Neuroscience of Vision." *Journal of Visual Art Practice*. 5(3): 195–213, 2006.

2-22. Arnheim, Rudolf. 1954. "Art and visual perception: A psychology of the creative eye." Univ of California Press: Oakland, California.

2-23. Kahneman, Daniel. 1973. "Attention and Effort." Prentice-Hall: Englewood Cliffs, New Jersey.

2-24. Vernon, Magdalen Dorothea. 2013. "A further study of visual perception." Cambridge University Press: Cambridge.

2-25. M. El-Sayed, J. El-Sayed. "Importance of Psychomotor Development for Innovation and Creativity." *International Journal of Process Education*. 4:89–94, 2012.

2-26. ISO. Technical Specification ISO/TS 16949. 2009. Quality management systems—Particular Requirements for the Application of ISO 9001:2008 for Automotive Production and Relevant Service Part Organizations, third edition. ISO: Geneva, Switzerland.

2-27. El-Sayed, M., "Rethinking the Automotive Design and Development Processes for Product Realization," SAE Technical Paper 2008-01-0861, 2008, doi:10.4271/2008-01-0861.

2-28. Siemens. Process planning tools support Volvo Cars' expansion to Asia Pacific region. Available from: http://www.plm.automation.siemens.com/CaseStudyWeb/dispatch/viewResource.html?resourceId=25856.

2-29. http://www.altairhyperworks.com/industry/Automotive.

2-30. El-Sayed, M., "Value Streaming Through Customer Participation in Product Realization," *SAE Int. J. Mater. Manf.* 9(3):552-557, 2016, doi:10.4271/2016-01-0344.

Chapter 3
Process Characteristics and Attributes

"To improve is to change; to be perfect is to change often."

—Winston Churchill

3.1 Introduction

Since the introduction of the automobile, the rate of change due to consumers' demand for improvements has been steadily increasing. With global competition, more stringent safety standards, and the quest for more environmentally friendly vehicles, this rate of change is being rapidly accelerated. To face these challenges, automotive companies must continuously strive to improve their vehicles and the processes by which these vehicles are developed and produced. To stay competitive, automotive companies must work on their development and production processes to improve quality, reduce cost, and minimize development time from concept to production.

As discussed in Chapter 1, the vision and mission upon which realization and production processes are built shape their attributes and ultimately determine the outcomes. To achieve perfection by design or through steady improvements, continuous changes in realization and production processes are inevitable. Therefore, adopting the proper strategies for creating better processes and managing improvements is crucial for continued success. With the proper strategy, any process could be improved as long as the desired process characteristics and attributes are well defined and correctly pursued. To guide the design and continuous improvement of realization processes, a framework for identifying desired process attributes, criteria, and targets is presented in this chapter.

3.2 Framework for Realization Process Characteristics and Attributes

As discussed in Chapter 2, any product realization process has three main phases: inception, conception, and maturation. Each phase by itself is a realization in a different domain. These domains are the perceptual, virtual, and physical domains, respectively. By the end of the maturation phase, both product and production process are physically realized and ready for mass production. While the realization process is usually completed by the realization of both the product and the mass production process, the mass production process continues as long as the product remains in production.

The activities of the realization cycle are usually concluded soon after the start of production, as illustrated by the dashed lines in Figure 3.1. Since the mass production process itself is an outcome of the realization process phases, both realization and production processes may share the same vision for the product. Accordingly, the main characteristics and attributes of both processes could have some broad similarities. Differences between the two processes come from the nature of each in addition to the implementation and operational targets.

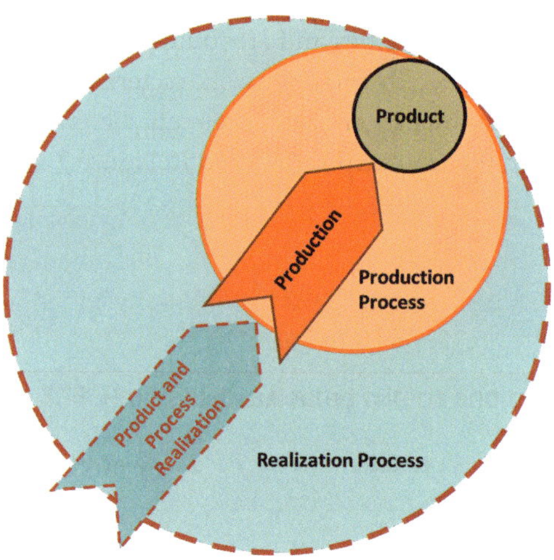

Figure 3.1 Development and mass production realization processes.

In the following, a general framework for process characteristics and attributes is established. The selected process generic characteristics and attributes, in the established frame work, are to some extent applicable to both realization and mass production processes.

Whether through evolution or by design, every product realization process has multi-characteristics and attributes that distinguish the process, the product, and, to some extent, the corporate culture. In general, it takes a set of criteria to shape an attribute. Similarly, it takes a set of attributes to establish a characteristic, as illustrated in Figure 3.2.

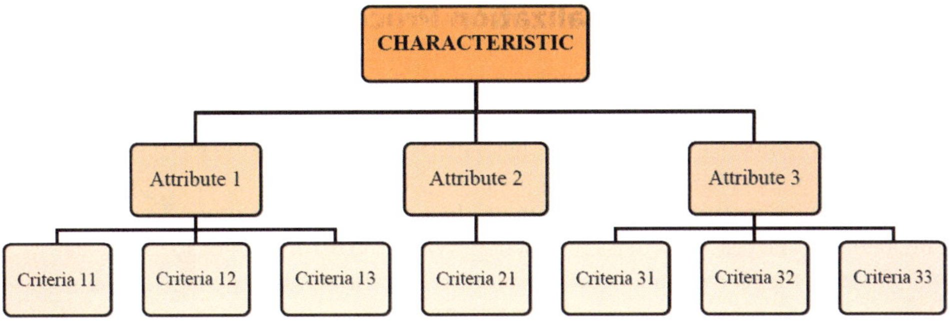

Figure 3.2 Process characteristics, attributes, and criteria.

Some of the basic process characteristics and attributes can be classified into two main groups:

- form characteristics and attributes
- function characteristics and attributes

The form characteristics and attributes are composed of the two subgroups of architecture and implementation characteristics and attributes. The function characteristics and attributes contain the process health and utility characteristics and attributes. For illustration, some examples of the major generic process characteristics and attributes in each subgroup are presented in the framework outlined in Figure 3.3.

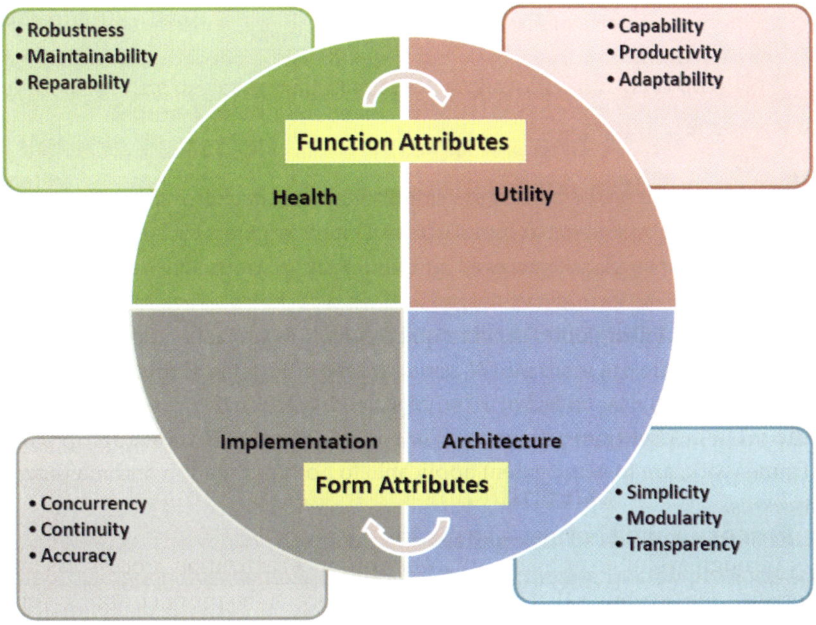

Figure 3.3 Framework for form and function process characteristics and attributes.

While Figure 3.3 represents the framework for the process characteristics and attributes, three generic attributes are selected as examples from each category. To realize any of the attributes shown in Figure 3.3, a set of criteria must be identified and specific targets must be established for each one of these criteria, as illustrated in Figure 3.4. The established set of targets should be specific, realizable, and measurable. The level of achievement for any desired process attribute depends on the established targets, as well as the level of achievement for each target.

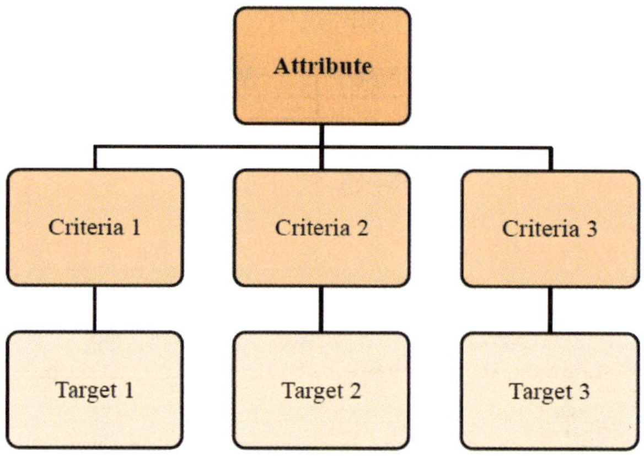

Figure 3.4 Transition from attributes to targets.

As shown in Figure 3.4, several criteria are usually involved in achieving each attribute. In this chapter, three criteria are selected as examples for each attribute. While one target is shown for each one of the criteria, several targets may be needed in some cases. In the following, the selected generic attributes, selected relevant criteria, and selected target for each one of the criteria will be discussed as examples.

3.3 Process Form Attributes

The set of attributes built in the process, through its architecture and desired implementation characteristics, are considered as the form attributes. Different processes have different forms by design or through evolution. Also, under pressure or slowly over time, the form attributes of a process may suffer undesirable drift or shift. In the following, as examples, three selected generic architectural attributes and three selected generic implementation attributes are considered. While there are other relevant attributes that could be included in each category, the selected examples are used to illustrate the framework presented in Figure 3.3.

3.3.1 Process Architecture Attributes

To illustrate the steps for achieving the realization process or production process desired architecture, three generic attributes are selected. For each attribute, three criteria and a target for each one of these criteria are used as examples.

3.3.1.1 Simplicity: One of the key attributes for process effectiveness in dealing with the competitive pressures and cost reduction is simplicity [3-1]. With increased demands for better quality at lower prices, the need for realization process and production process simplicity is higher. Simplicity can be achieved by eliminating or reducing complexity in all process phases, subphases, and tasks. Focusing on the architectural aspect of simplicity, as shown in Figure 3.5, the following three criteria can be selected:

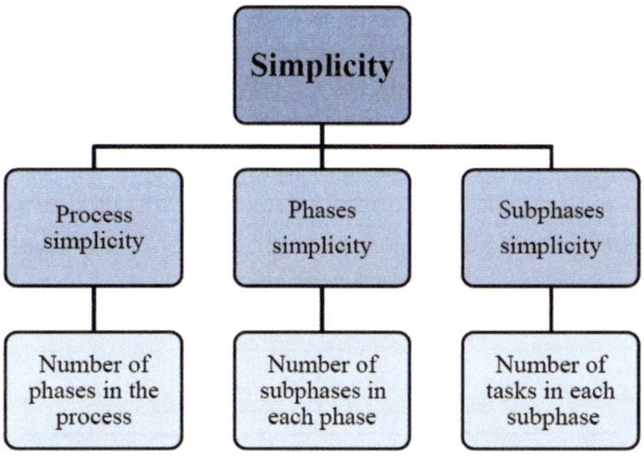

Figure 3.5 Simplicity selected criteria and targets.

- process simplicity
- phases simplicity
- subphases simplicity

For these three criteria, the following realistic and measurable targets are selected, as shown in Figure 3.5:

- number of phases in the process
- number of subphases in each phase
- number of tasks in each subphase

It is evident that process complexity would be reduced when the corresponding target for any of the three criteria is decreased. By continuously reducing the targets and achieving them, simplicity could be improved. Ultimate simplicity would be realized when the attempt to eliminate or combine any phases, subphases, or tasks would not succeed without the potential of causing an adverse effect on the process, product, or the people involved. To illustrate, the following example is considered.

3.3.1.1.1 Example 3.1: Considering the realization process discussed in Chapter 2, two of the simplicity criteria and targets can be easily observed:

- number of phases = 3
- number of subphases in each phase = 3

By keeping the number of tasks needed to achieve desired outcomes in each subphase to a minimum, realization process simplicity can be achieved.

3.3.1.2 Modularity: The elements of a modular process are usually grouped into modules following a logical architecture. Modularization generally allows for some of the activities to be performed in parallel with the other modules. In addition, within a modular architecture, some elements could be easily altered or replaced, with no effect on the other modules [3-2].

By design or through continuous improvements, a considerable level of modularity could be achieved in realization or production process phases, subphases, and tasks. Focusing on the architectural aspect of modularity, as shown in Figure 3.6, the following three criteria can be selected:

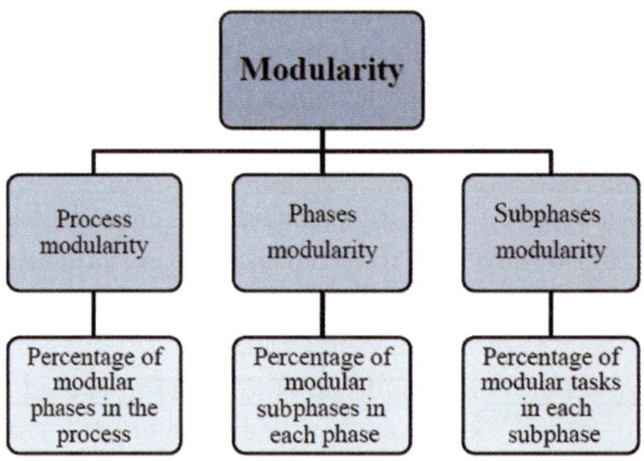

Figure 3.6 Modularity selected criteria and targets.

- process modularity
- phases modularity
- subphases modularity

Considering all possible forms of modularization, the following targets are selected, as shown in Figure 3.6:

- percentage of modular phases in the process
- percentage of modular subphases in each phase
- percentage of modular tasks in each subphase

Process modularity would be improved when a higher percentage is targeted and achieved for any of the three criteria. Ultimate modularity would be achieved when the attempt to increase any of the targets would not succeed without causing an adverse effect.

3.3.1.2.1 Example 3.2: Considering the realization process discussed in Chapter 2, two of the process modularity criteria and targets can be easily observed.

By implementing the realization phases in parallel, the percentage of modular phases in the process = 100.

With sequential implementation of inception, conception, and maturation subphases, percentage of modular subphases in each phase = 0.

By increasing the percentage of parallel tasks in each subphase, subphase modularity can be improved.

3.3.1.3 Transparency: A well-architected process helps the individuals involved with it to obtain a comprehensive view of its entirety [3-3]. Knowing the effect and significance of the work one is performing on the process and product is the most important aspect of a transparent process. In addition to continuous training of individuals, a considerable level of transparency could be achieved in process phases, subphases, and tasks through design and continuous improvements. Focusing on the architectural aspect of transparency, as shown in Figure 3.7, the following three criteria can be selected:

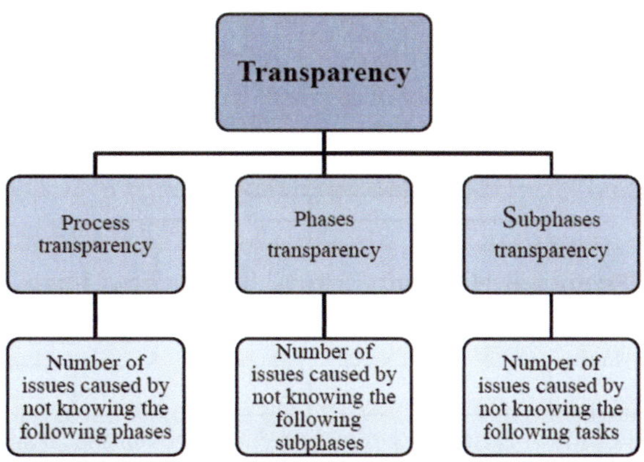

Figure 3.7 Transparency selected criteria and targets.

- process transparency
- phases transparency
- subphases transparency

For these three criteria, the following targets are selected, as shown in Figure 3.7:

- number of issues caused by not knowing the following phases
- number of issues caused by not knowing the following subphases
- number of issues caused by not knowing the following tasks

Clearly, process transparency would be improved when a lower number is targeted and achieved for any or all three criteria. Ultimate transparency would be realized when all the targets are set to zero and achieved.

3.3.2 Process Implementation Attributes

Three generic attributes are selected to illustrate the steps for process implementation. Three criteria and one target for each are selected as examples.

3.3.2.1 Concurrency: The success of a process in minimizing time and cost depends on the level of concurrency in the process, each individual phase, and each subphase. For example, while some concurrency between design and manufacturing can be fashioned during late development realization phases, the concurrency between marketing, planning, styling, and engineering particularly at early phases is harder to achieve. Achieving high level of concurrency, however, is necessary for ultimate one directional flow throughout the process.

By design or through continuous improvements, a substantial level of concurrency could be more easily achieved in process phases, subphases, and tasks. Focusing on the implementation aspect of concurrency, as shown in Figure 3.8, the following three criteria can be selected:

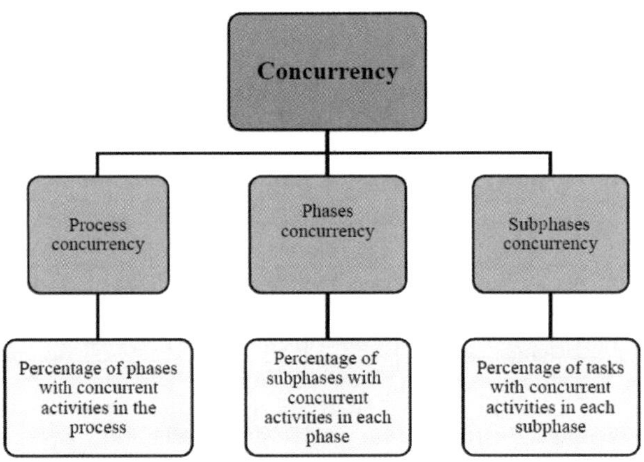

Figure 3.8 Concurrency selected criteria and targets.

- process concurrency
- phases concurrency
- subphases concurrency

Considering all possible concurrency between activities, the following targets are selected, as shown in Figure 3.8:

- percentage of phases with concurrent activities in the process
- percentage of subphases with concurrent activities in each phase
- percentage of tasks with concurrent activities in each subphase

With higher targets a better level of concurrency could be achieved. It is always easier and cheaper to resolve issues with more concurrency at early phases. Late design changes, after specifications have been frozen and tools are ordered, can be very costly and time-consuming. Through concurrency, information gathering and problem solving activities at the early phases, the development of product and production process can become faster and smoother [3-4]. It is also relatively less expensive to work concurrently on alternative concepts, during the early phases, where the most favorable concept can be easily identified [3-5], [3-6], [3-7].

3.3.2.2 Continuity: To provide continuity and smooth process flow during implementation, process phases should be well integrated with seamless transition. This integration should start from the front end to the field support backend until the full process "beginning-to-end" scope is realized [3-8]. Continuity could be improved by reducing the number of cycles, identifying problems early, and speeding problem solving cycles [3-5], [3-9]. Focusing on the integration and seamless transition, as shown in Figure 3.9, the following three criteria can be selected:

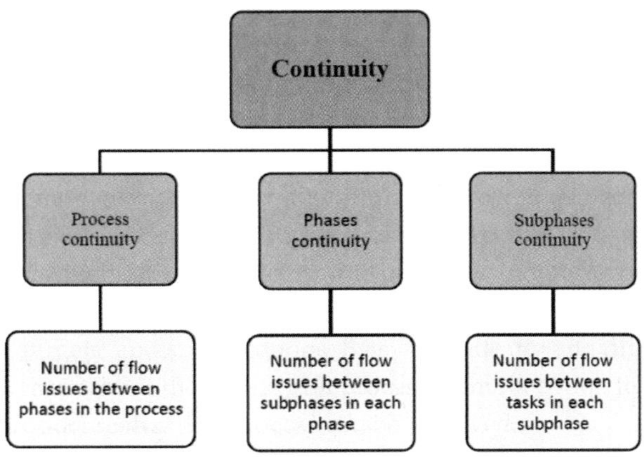

Figure 3.9 Continuity selected criteria and targets.

- process continuity
- phases continuity
- subphases continuity

For continuity and smooth flow throughout the process, the following targets are selected, as shown in Figure 3.9:

- number of flow issues between phases in the process
- number of flow issues between subphases in each phase
- number of flow issues between tasks in each subphase

Process continuity would be improved when a lower number are achieved for any of the three targets. Ultimate integration and seamless transition would be realized when all the targets are set to zero and achieved.

3.3.2.3 Accuracy: Performing accurate work on time is crucial for successful implementation of any process. In addition, the ability to precisely measure the process and product key parameters for assessing the gaps from established targets is vital for achieving desired process and product attributes. In general, accuracy in work, timing, and measurements can be improved by setting the right targets. Focusing on the implementation accuracy, as shown in Figure 3.10, the following three criteria can be selected:

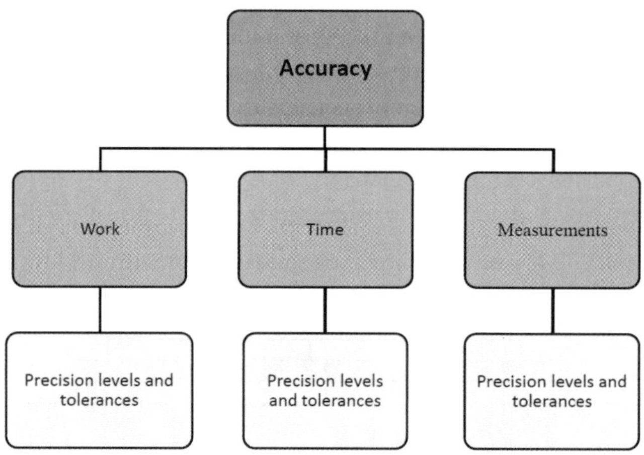

Figure 3.10 Measurability selected criteria and targets.

- work
- time
- measurements

For these implementation accuracy criteria, as shown in Figure 3.10, the following set of targets are selected examples:

- work precision levels and tolerances
- time precision levels and tolerances
- measurements precision levels and tolerances

With higher precision targets and tight tolerances, better process accuracy could be achieved. Better implementation accuracy would produce desired improvements in both process and product attributes.

3.4 Process Function Attributes

The set of attributes related to the process performance are considered as the function attributes. Different processes have different functions or performance by design or through evolution. Also, under pressure the same process may function or perform differently. In the following, as examples, three selected generic health attributes and three selected generic utility attributes are discussed. The health attributes are considered part of the function attributes because of the total dependence of the utility attributes on the health function for performance. While there are other relevant performance attributes that could be included in each of the two categories, the selected examples are used to illustrate the framework presented in Figure 3.3.

3.4.1 Process Health Attributes

3.4.1.1 Robustness: The ability of a process to withstand variations in the operational environment is dependent on its robustness. Therefore, robustness of any process can be characterized by its tolerance to unexpected variations and events. Focusing on the robustness in facing variations, as shown in Figure 3.11, the following three criteria can be selected:

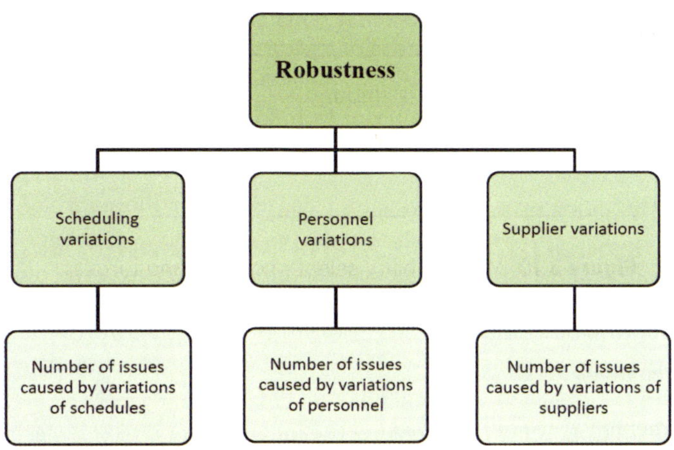

Figure 3.11 Robustness selected criteria and targets.

- scheduling variations
- personnel variations
- supplier variations

For these criteria, the following targets are selected, as shown in Figure 3.11:

- number of issues caused by variations of schedules
- number of issues caused by variations of personnel
- number of issues caused by variations of suppliers

With the reduction in the number of issues targeted and achieved for all three criteria, process robustness could be improved. Ultimate robustness would be realized when all the targets are set to zero and achieved.

3.4.1.2 Maintainability: Process maintenance or upkeep is important for healthy performance and delivery of expected results during operations. In addition to the upkeep under normal operations, maintainability of the process may be performed to adjust for product or operational changes. Focusing on maintainability due to changes, as shown in Figure 3.12, the following three criteria can be selected:

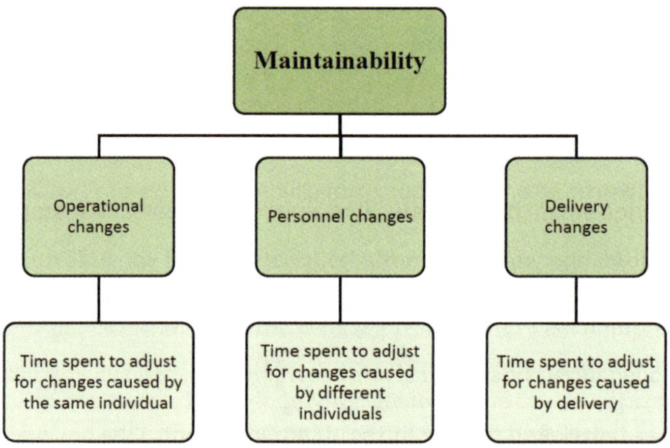

Figure 3.12 Maintainability selected criteria and targets.

- operational changes
- personnel changes
- delivery changes

Focusing on the time aspect of these criteria the following targets are selected, as shown in Figure 3.12:

- time spent to adjust for changes caused by the same individual
- time spent to adjust for changes caused by different individuals
- time spent to adjust for changes caused by delivery

With lower targeted and achieved times for all three criteria, process maintainability would be improved. With continuous improvement, minimum process maintainability and upkeep time could be achieved.

3.4.1.3 Reparability: Process repair is sometimes necessary for bringing the process back to the healthy performance level expected for product development. During the scheduled or unexpected process repairs, downtime and cost due to failures are unavoidable. Focusing on the repairs due to failures, as shown in Figure 3.13, the following three criteria can be selected:

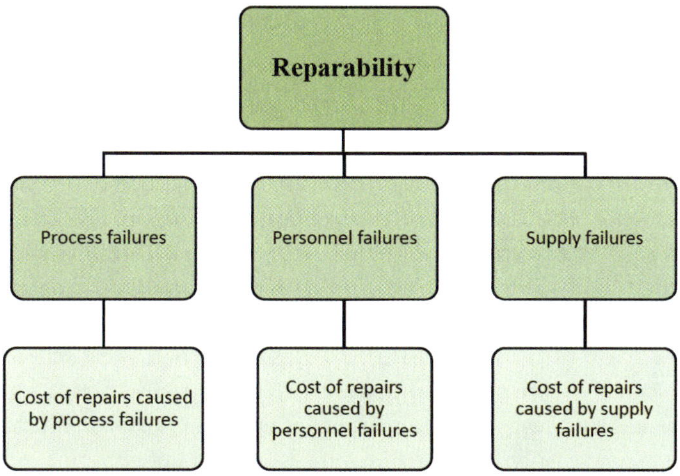

Figure 3.13 Reparability selected criteria and targets.

- Process failures
- Personnel failures
- Supply failures

Focusing on the cost aspect of these criteria, the following targets are selected, as shown in Figure 3.13:

- Cost of repairs caused by process failures
- Cost of repairs caused by personnel failures
- Cost of repairs caused by supply failures

With lower achieved targets for the costs due to failures, a better process reparability could be achieved. Better reparability would reduce process costs and improve profitability.

3.4.2 Process Utility Attributes

Utility attributes of the development and production processes are usually different. The following selected generic attributes are more geared toward the production process, but to some extent are still applicable to the development process. Some of the differences in setting operational targets and implementation will be discussed.

3.4.2.1 Capability: In general, process capability includes current and potential abilities to deliver desired outcomes. For the realization process as a whole, these abilities include the following:

- capable of achieving development process quality, timing, and cost targets
- capable of achieving production process quality, timing, and production targets
- capable of achieving organizational strategic, current business, and customer satisfaction goals

Accordingly, realization process capability could be measured by the realized production process capability to achieve production process quality, timing, and production targets. Focusing on quality, a production process capability can be defined as the ability of the process to meet established product specifications. This process capability is therefore one of the key utility attributes for reducing production variability and achieving targeted product quality. To measure production process capability, the process output is compared with the specification limits. To perform the comparison, the process must be kept in statistical control by taking the appropriate and necessary actions using statistical methods to analyze the process or its outputs [3-10]. Process parameters, also called "process characteristics," especially the dominant parameters that significantly impact product quality, are measured. In addition, process spread under normal operation is quantified. Focusing on achieving quality, the three criteria shown in Figure 3.14 are selected for production process capability:

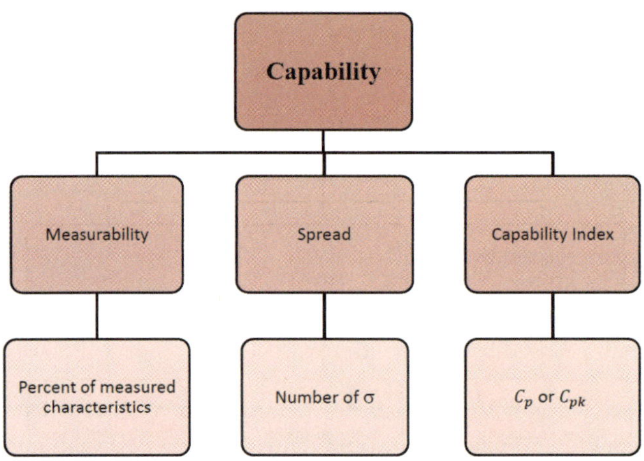

Figure 3.14 Suitability selected criteria and targets.

- measurability
- spread
- capability index

Considering some possible capability aspects, the following targets are selected, as shown in Figure 3.14:

- percent of measured characteristics, especially dominant characteristics
- number of σ. For example, $\pm 3\sigma$ could be used for 6σ spread
- C_p or C_{pk} (established target must be greater than 1, as will be discussed in later chapters)

Higher process capability would be realized when close to 100% of process parameters are measured, addressed, and process is kept in statistical control. With higher numbers of σ (standard deviation), process performance and product quality could be improved. With a higher capability index, the process spread could be contained within the specified tolerance, reducing the percentage of out-of-spec or faulty products.

It is clear from the selected criteria that process capability is a vital utility attribute for the production process. As mentioned earlier, for the realization process, in addition to its many unique capability attributes, some of its capability attributes could be directly related to the capability attributes of the production process it produces.

3.4.2.2 Productivity: Process productivity is an essential utility and performance attribute for corporate vitality and competitiveness. To stay competitive realization and production processes must be designed, developed, and continuously improved with specified productivity targets for pertinent criteria. Focusing on competitive aspects of productivity, as shown in Figure 3.15, the following three criteria could be selected:

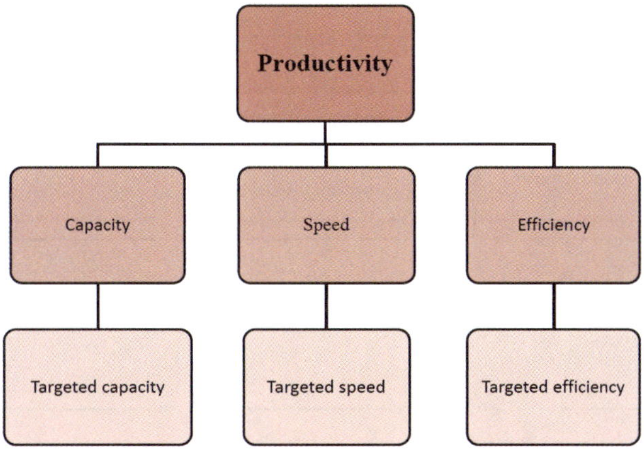

Figure 3.15 Productivity selected criteria and targets.

- capacity
- speed
- efficiency

Considering these significant productivity criteria, the following targets are selected, as shown in Figure 3.15:

- targeted capacity
- targeted speed
- targeted efficiency

Designing and developing for targeted productivity criteria are necessary for production process utility. For realization process, in addition to many unique productivity attributes that can be considered, its productivity attributes could be directly related to the productivity attributes of the production process it produces.

3.4.2.3 Adaptability: With increased pressures of rapidly changing technologies, and continuously changing markets, process adaptability becomes one of the most sought after utility process attributes. To design, develop, and achieve adaptability, as shown in Figure 3.16, the following three criteria examples are selected:

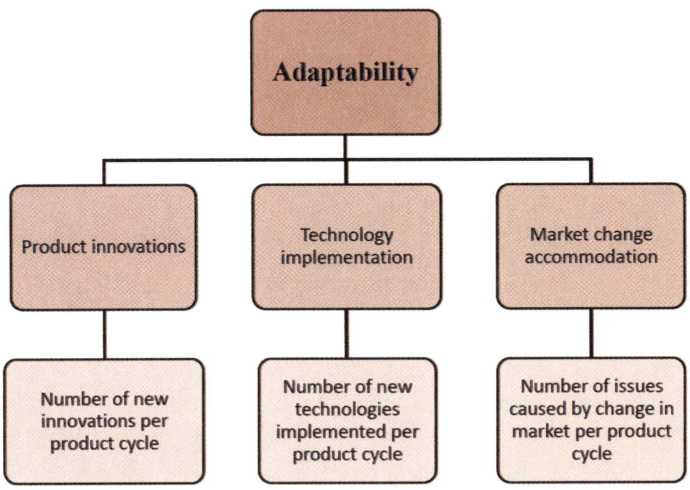

Figure 3.16 Adaptability selected criteria and targets.

- product innovations
- technology implementation
- market change accommodation

For these three criteria, the following targets are selected, as shown in Figure 3.16:

- number of new innovations per product cycle
- number of new technologies implemented per product cycle
- number of issues caused by change in market per product cycle

These adaptability criteria and targets are applicable for utility of both realization and production processes.

For realization process, more innovations and new opportunities can be identified at the early realization phases known as the fuzzy front end [3-11]. With higher targeted innovations and new technology implementation, in addition to a lower number of issues due to market changes, a better level of adaptability could be achieved. To face uncertainty and changing markets during production, achieving adaptability attributes and criteria is necessary for the production process.

3.5 Process Fit Attributes

In addition to form and function process characteristics and attributes, there is the essential set of fit characteristics and attributes. The set of fit characteristics and attributes depicts the degree of achievement and level of fitness in meeting intended purpose. Therefore, this set of characteristics and attributes is more summative in nature. It covers characteristics and attributes that transcend both the form and function sets. Focusing on some key aspects of process fit, as shown in Figure 3.17, the following three attributes are selected:

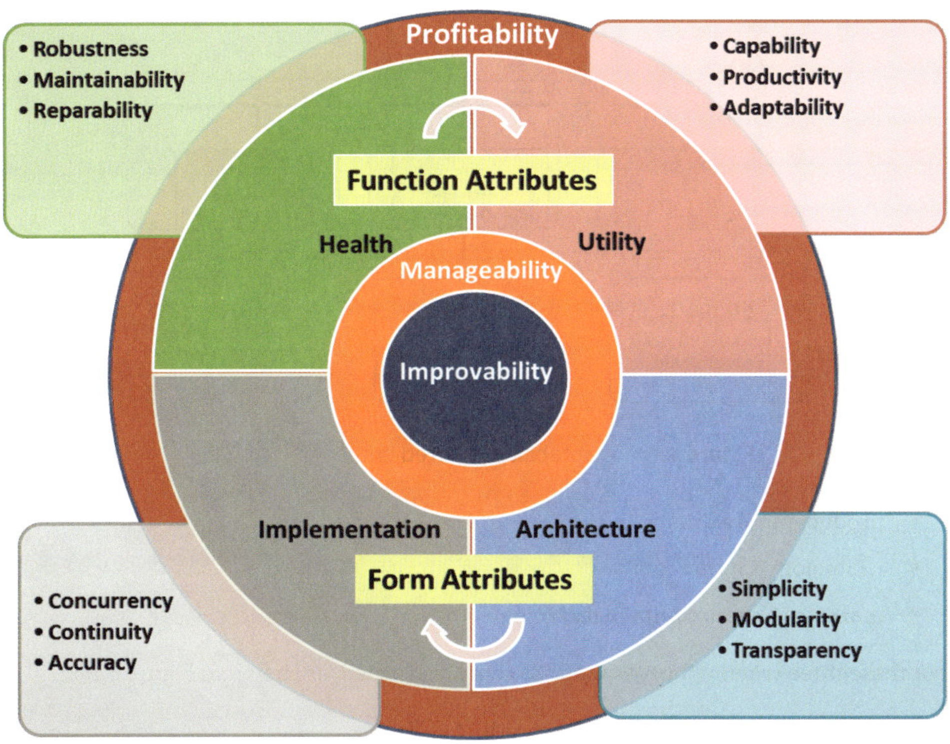

Figure 3.17 Fit, form, and function process characteristics and attributes.

- profitability
- manageability
- improvability

These three key fit attributes are selected to illustrate the steps for achieving the desired fit for both realization and production processes. For each attribute, three criteria and a target for each criterion are used as examples.

3.5.1 Profitability

Profitability is an essential process fit attribute. For sustainable growth, successful realization and production processes must be profitable. To achieve corporate objectives and business profitability goals, the following three criteria, shown in Figure 3.18, are selected to illustrate the concept:

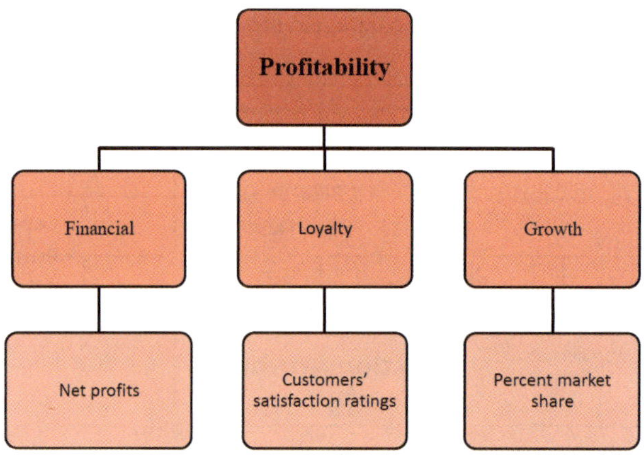

Figure 3.18 Profitability selected criteria and targets.

- financial
- loyalty
- growth

For these three criteria, the following targets are selected examples, as shown in Figure 3.18:

- net profits
- customers' satisfaction ratings
- percent market share

With higher net profit, better business and financial status could be achieved. For continual growth in customers' satisfaction ratings, loyalty, and market share, better corporate stability could also be achieved. Therefore, achieving these criteria is necessary for continued business success and corporate prosperity. The discussed profitability criteria and targets are equally applicable to both development and production processes.

3.5.2 Manageability

Manageability is the ability to manage all aspects of the process and is a key fit attribute for process and business success. Usually, there is a strong connection between any realization or production process attributes and the corporate structure. A full understanding of this connection can assist in creating a better quality product. In addition, the efficiency of the personnel interactions required in implementing the process can be enhanced [3-12]. To demonstrate process design and development for manageability, the following three criteria, shown in Figure 3.19, are selected as examples:

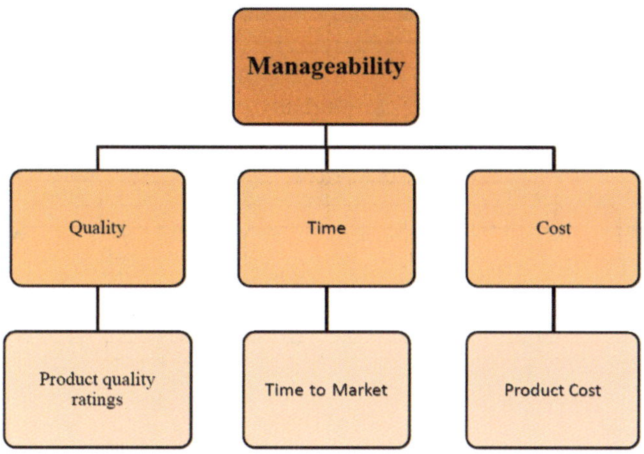

Figure 3.19 Manageability selected criteria and targets.

- quality
- time
- cost

For these three criteria, the following targets are selected examples, as shown in Figure 3.19:

- product quality ratings
- time to market
- product cost

The result of achieving these manageability targets is bringing quality products to market faster and cheaper. Some reported efforts show that better manageability in conjunction with other process attributes have resulted in measurable improvements in time to market and product quality [3-8]. Therefore, these selected manageability criteria are necessary for both development and production processes.

3.5.3 Improvability

Process improvability is directly connected to product quality. Since quality is usually achieved through continuous improvement, improvability was chosen as the peak and central summative attribute that provides the ability to improve all other process attributes. To demonstrate process design and development for improvability, the following three criteria, shown in Figure 3.20, are selected as examples:

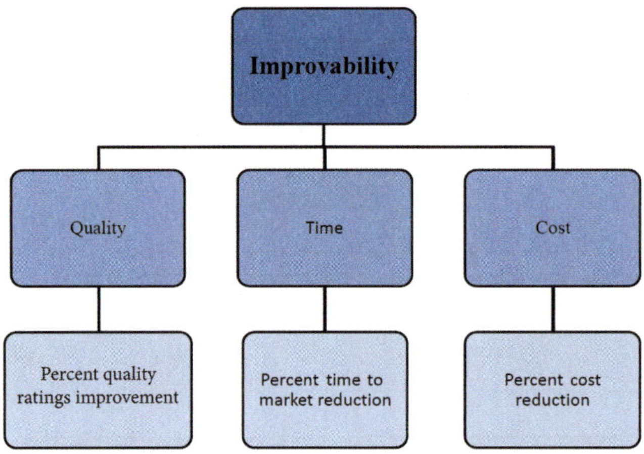

Figure 3.20 Profitability selected criteria and targets.

- quality
- time
- cost

For these three criteria, the following targets are selected examples, as shown in Figure 3.20:

- percent quality rating improvement
- percent time to market reduction
- percent cost reduction

These process improvability criteria and targets provide a framework for improvability of both development and production processes by design or through continuous improvements. Process improvements, however, should be built in with closed-loop mechanisms [3-8], [3-13]. Closing the loop can be achieved by linking product improvement requirements with process improvement initiatives. These initiatives or corrective actions should be driven to close process and product performance gaps. To facilitate continuous improvements, the following additional process elements should be included [3-8]:

- checkpoints for each phase and subphase
- quick decision making procedures
- monitoring and upgrading skills of all involved individuals

With similar selected characteristics, as shown in Figure 3.19 and Figure 3.20, the ultimate targets for process manageability and process improvability are the production of quality product—faster and cheaper. Accordingly, process improvability could be joined with process manageability, as shown in Figure 3.21. The resulting joint process manageability and improvability fit characteristic will be discussed further in Chapter 5.

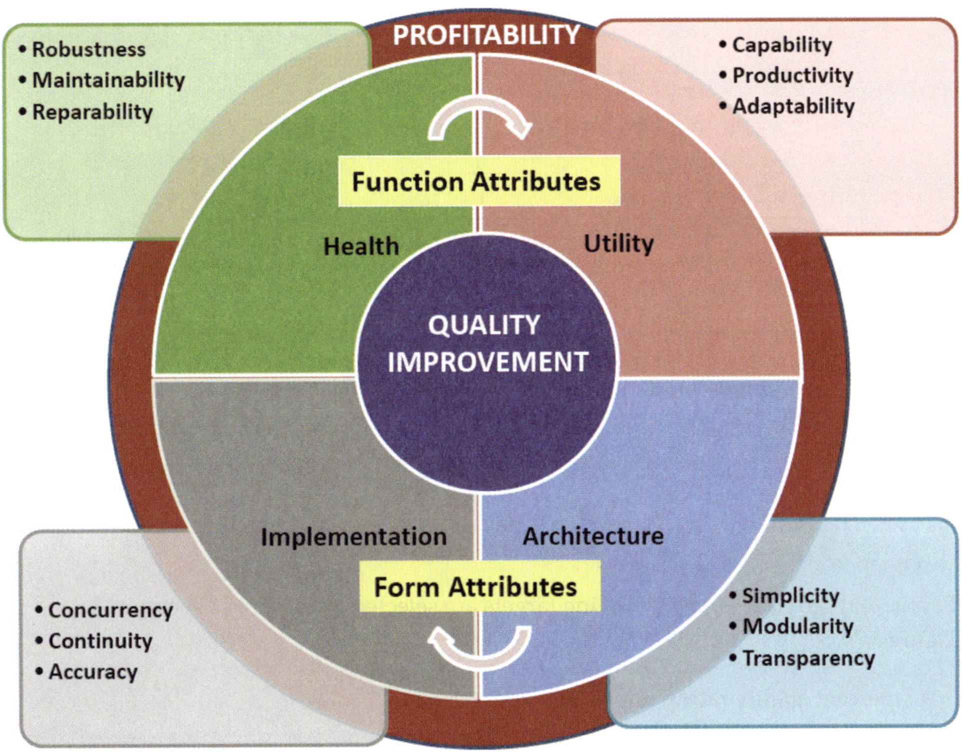

Figure 3.21 Joined manageability and quality fit characteristic.

3.6 Learning Outcomes from Process Characteristics and Attributes

The desired learning outcomes from studying process characteristics and attributes can be categorized as follows.

3.6.1 Knowledge

Process characteristics and attributes can be classified into three sets: form, function, and fit characteristics and attributes. The form and the function sets are composed of two subsets. The set of form characteristics and attributes is composed of the architecture and the implementation subsets. The set of function characteristics and attributes is composed of the health and utility subsets. The fit set is composed of the process summative characteristics and attributes.

Designing a process with desired set of characteristics and attributes may take several criteria and attributes to build a process character. In general, any meaningful and sustainable process attribute must be built at the criteria level. Therefore, identifying the set of criteria shaping an attribute is a key step toward achieving the desired attribute. Any criterion should be absolutely or relatively measurable. Once a criterion is identified, realistic target or targets must be established and pursued. During realization, the gap from any established target should be utilized in magnitude and direction to determine the required adjustment or improvement for the specific criterion.

The main objective during designing and developing any realization process is to achieve all desired characteristics and attributes. However, achieving perfection in a process during initial design may not be easy or possible. Also, under operational pressures, process attributes may become shifted or drifted from original intent. Therefore, directing attention to process improvability during design and during repeated operational cycles, such as the case with mass production of vehicles, is essential. The best course of action is to include improvability attributes and criteria during the inception of any process.

For process manageability and improvability, adopting the proper strategy is necessary. Any proper strategy must start with identifying the set of desired attributes and systematically pursue their improvements or maintenance. This may also include identifying the set of undesired process attributes and systematically pursuing their elimination or reduction.

3.6.2 Skills

The skills needed for managing process design, development, and improvements depend on the ability to identify the set of desired process attributes and systematically pursue their achievement or improvement. This also requires building the skills of identifying the correct criteria and establishing realistic targets. Success during implementation, however, hinges on the ability to measure the gaps for established targets and using the magnitudes and directions for making the necessary corrections.

3.6.3 Paradigm/Attitude

Process design and development starts with identifying desired characteristics and attributes. For mass production, the characteristics and attributes of both realization and production processes shape product characteristics and attributes. Along with profitability, process manageability and improvability fit attributes are essential for continued success.

3.7 References

3-1. El-Sayed, M., "Rethinking the Automotive Design and Development Processes for Product Realization," SAE Technical Paper 2008-01-0861, 2008, doi:10.4271/2008-01-0861.

3-2. Baldwin, C. Y., Clark, K. B., "Modularity in the Design of Complex Engineering Systems." Complex Engineered Systems: Science Meets Technology. Ali Minai, Dan Braha, and Yaneer Bar Yam (Eds). New England Complex Systems Institute Series on Complexity. Springer-Verlag: New York, 2006.

3-3. Wheelwright, S. C. and K. B. Clark. 1992. "Revolutionizing Product Development, Quantum Leaps in Speed, Efficiency and Quality." The Free Press: New York.

3-4. Thomke, S., and Fujimoto T. "The Effect of 'Front-Loading' Problem-Solving on Product Development Performance," *Journal of Product Innovation Management.* 17(2):128–142, 2000.

3-5. Svensén, U., Beskow, C., Hagås, K., Heper, Y., Nilsson, K., and Söderquist, K., "Global Study on Product Development." ENRIA: Stockholm, 2000.

3-6. Sobek, D. K., II, Liker, J. K., Ward, A. C. "Another Look at How Toyota Integrates Product Development," Harvard Business Review, July to August 1998.

3-7. Lawrence, P. R., and Lorsch J. W. 1967. "Organization and Environment: Managing Differentiation and Integration." Harvard Business School Classics. Harvard Business School Press: Boston.

3-8. Holmes, M. F., and Campbell, R.B. 2003. "Product Development Processes: Three Vectors of Improvement." Center for Innovation in Product Development, MIT: Massachusetts.

3-9. Montgomery, D. 2012. "Statistical Quality Control," 7th edition, Wiley: New Jersey.

3-10. Koen, P.A., Ajamian, G.M., Boyce, S., Clamen, A., Fisher, E., Fountoulakis, S., Johnson, A., Puri, P., and Seibert, R." Fuzzy Front End: Effective Methods, Tools and Techniques." The PDMA Tool book for New Product Development, Belliveau, P., Griffin, A., and Somermeyer, S (Eds). Product Development and Management Association: Chicago. 2002.

3-11. Gulati, R., K., and Eppinger, S., D., "The coupling of product architecture and organizational structure decisions," Working Paper No. 151, MIT International Center for Research on the Management of Technology: Cambridge, MA. 1996.

3-12. Patterson, M. L. 1993. "Accelerating Innovation," Van Nostrand Reinhold: New York.

3-13. Chrysler Corporation, Ford Motor Company, and General Motors Corporation, *Advanced Product Quality Planning (APQP) and Control Plan Reference Manual*, June 1994.

Chapter 4
Vehicle Characteristics and Attributes

"Quality is everyone's responsibility."

—W. Edwards Deming

4.1 Introduction

Product realization is the core of engineering. In fact, engineering can be defined as the art and science of product realization. Product realization embodies all activities that deal with bringing products to being such as research, design, development, and problem solving. These realization activities also include planning, manufacturing, marketing, selling, maintenance, and repair. Similar to process realization discussed in the previous chapter, the vision upon which a product is realized shapes its characteristics and attributes, which ultimately determine its utility and performance. Achieving a desired product by design requires identifying the desired sets of characteristics, attributes, criteria, targets, and specifications. By systematically and diligently pursuing these desired sets, throughout the realization process from inception to production, the objectives of the product realization process can be achieved.

Since the introduction of the automobile, the pressures for improvements have been propelled by consumers' demand for innovation, efficiency, safety, and quality. Consequently, identifying the desired characteristics, attributes, and criteria in addition to setting the right targets are vital steps for developing the desired final vehicle and its technical specifications. Using technical specifications of previously developed vehicles as guide can enhance and facilitate the development of the new specifications. However, using technical specifications of previously developed vehicles, by altering few entries, may not lead to the development of the new desired vehicle attributes and characteristics. Therefore, adopting and implementing the proper methodology for developing the desired Vehicle Technical Specification (VTS) is essential. Any proper methodology, however, can be enabled by a well-defined vehicle characteristics framework [4-1]. In this chapter, a framework for developing vehicle characteristics, attributes, and criteria to facilitate the development of desired vehicle specifications and guide the vehicle realization process is outlined.

4.2 Framework for Product Characteristics and Attributes

Every product has several key characteristics and attributes that define the product and shape its usage. In general, it takes a set of criteria to shape an attribute and a set of attributes to establish a character or a product characteristic, as shown in Figure 4.1.

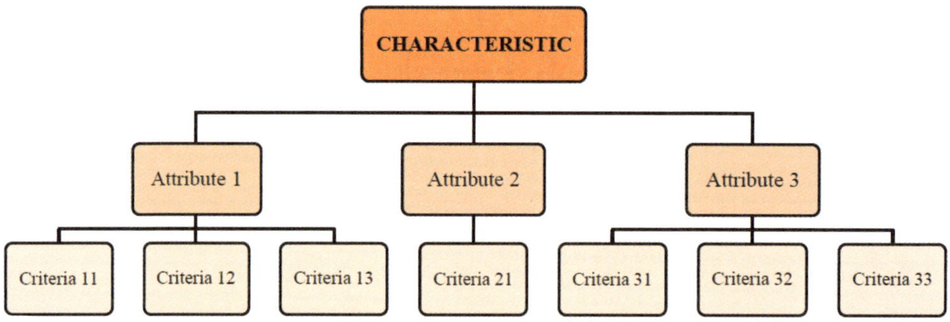

Figure 4.1 Product's characteristics, attributes, and criteria.

Some of the product characteristics and attributes can be classified into two main groups: the form and the function characteristics and attributes [4-2]. The form characteristics and attributes are composed of the two subgroups of architecture and manufacturing characteristics and attributes. The function characteristics and attributes contain the product health and utility characteristics and attributes. The framework outlined in Figure 4.2 shows some examples of the major vehicle characteristics in each subgroup.

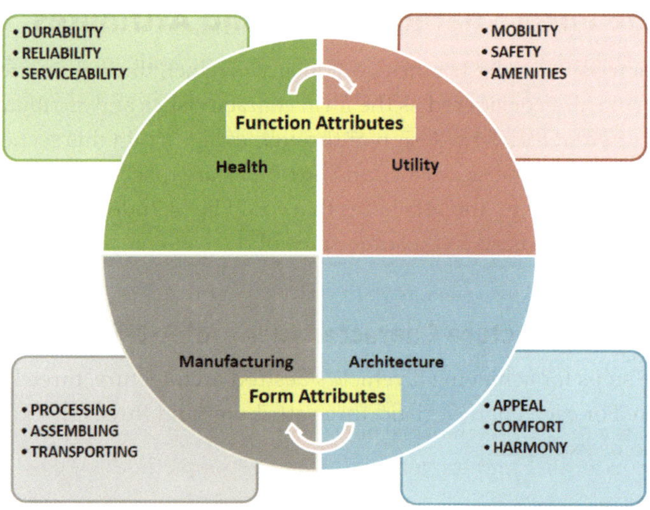

Figure 4.2 Framework for vehicle form and function characteristics.

For any attribute to be realized, the set of criteria forming the attribute must be identified and specific targets must be set for each one of these criteria, as shown in Figure 4.3. The set of targets should be specific, attainable, and measurable. The level of achievement, for any desired product attribute, and in turn characteristic, depends on the set of targets as well as the level of achievement for each.

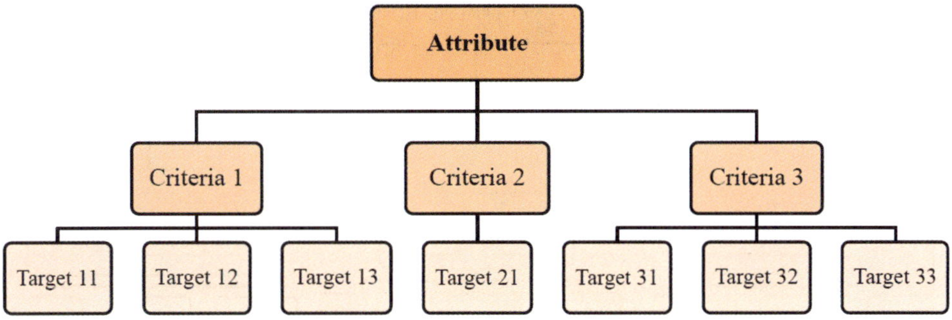

Figure 4.3 Attributes, criteria, and targets.

In the following, some form and function characteristics and attributes are selected, as examples, from each subgroup presented in the framework outlined in Figure 4.2. Also, relevant criteria for some of these attributes are identified and associated targets are discussed as examples.

While one specified target may be sufficient to achieve a specific criterion, several targets may be needed in some cases. The vehicle-level targets established for different criteria constitute the basis for the vehicle technical specifications (VTS). After the full vehicle VTS are developed, the subsystem technical specifications and the component technical specifications (CTS) can be derived.

4.3 Vehicle Form Characteristics and Attributes

The set of characteristics and attributes built in any product, through its architecture and manufacturing, are considered as the form characteristics and attributes. Different products have different forms by design or through evolution. In this section, some examples of vehicle form characteristics and attributes are discussed. While there are other relevant characteristics and attributes that could be included in each category, the selected examples are used to illustrate some of the contents for the framework presented in Figure 4.2.

4.3.1 Vehicle Architecture Characteristics and Attributes

To illustrate the steps for achieving a vehicle's desired architecture, three key characteristics are chosen. For each characteristic, three attributes and three criteria for each attribute are selected as examples.

4.3.1.1 Appeal: The first perceptually received vehicle impression and initial receptive realization builder. For example, the exterior appeal resulting from vehicle styling determines the vehicle attractiveness and aesthetics [4-3]. Focusing on the exterior appeal, as shown in Figure 4.4, three sets of attributes are selected as examples:

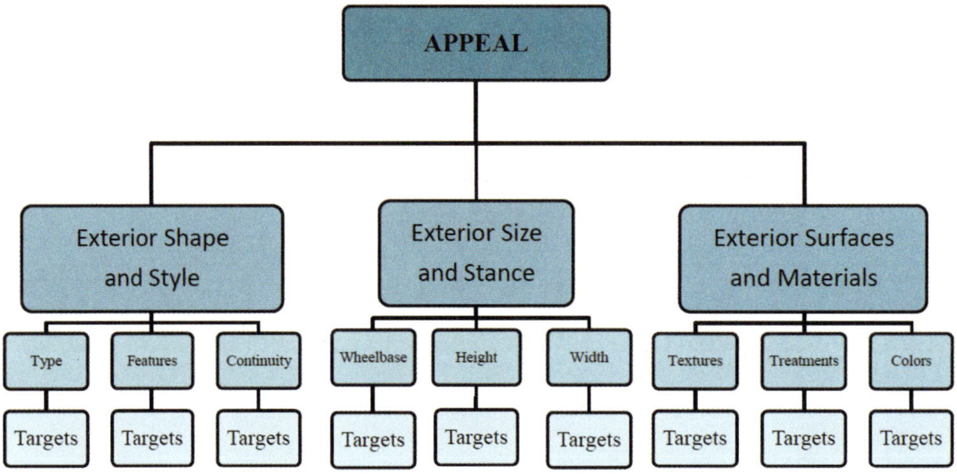

Figure 4.4 Selected attributes and criteria based on the appeal.

- *Exterior Shape and Style:* Delineate vehicle attractiveness and desirability
- *Exterior Size and Stance:* Determine vehicle presence and visual impressions
- *Exterior Surfaces and Materials:* Express vehicle finish and permanency

For exterior shape and style, the following three criteria can be selected as examples:

- *Type:* Targets for this criterion could be mid-size sedan, station wagon, minivan, sport utility vehicle, etc. [4-4].
- *Features:* Targets include topological and artistic particulars following the targeted vehicle theme such as classical, contemporary, or futuristic.
- *Continuity:* Targets include the level of continuity among vehicle surfaces [4-5].

For exterior size and stance, the following three criteria can be selected as examples:

- *Wheelbase:* Target for this criterion is vehicle wheelbase dimension.
- *Height:* Target for this criterion is vehicle height dimension.
- *Width:* Target for this criterion is vehicle width dimension.

For exterior surfaces and materials, the following three sets of criteria can be selected as examples:

- *Textures:* Targets include materials finish and surface properties.
- *Treatments:* Targets include necessary surfaces and material treatment such as paint or polish.
- *Colors:* Targets include surfaces and materials colors, tones, shine, and sparkle, etc.

Due to its artistic and visual nature, most styling and exterior appeal targets are usually easier to specify graphically using vehicle renderings, using virtual reality, using physical models, or through a combination using augmented reality. Accordingly, these subjective targets could be expressed using specified levels of approval based on evaluations by customers and experts.

4.3.1.2 Comfort: Encompasses vehicle amenities and accommodations. For example, as the most and closest interface area, vehicle interior sets the mood for vehicle comfort, coziness, and pleasability. Focusing on interior comfort, as shown in Figure 4.5, three sets of attributes are selected as examples.

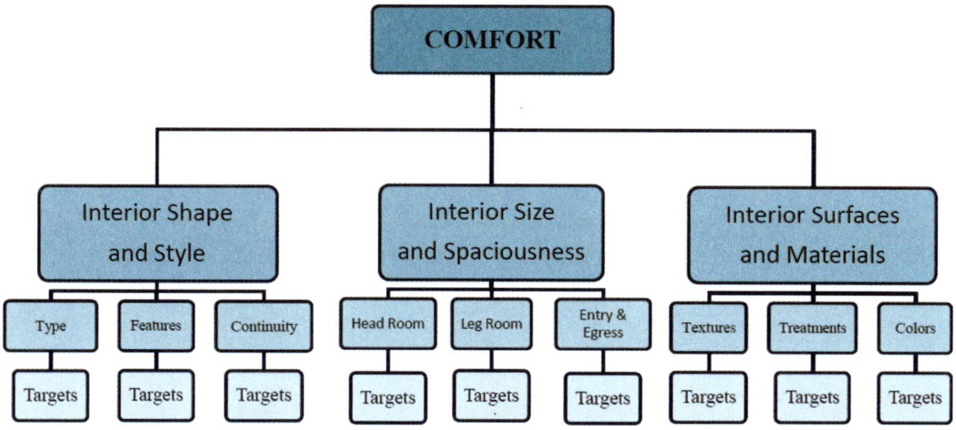

Figure 4.5 Selected attributes and criteria based on comfort.

- *Interior Shape and Style:* Portrays vehicle interior inevitability and coziness
- *Interior Size and Spaciousness:* Embodies all interior extents and roominess elements
- *Interior Surfaces and Materials:* Expresses interior build, finish, and longevity

For interior shape, the following three criteria can be selected as examples:

- *Type:* Targets for this criterion could be family car interior, sporty car interior, luxury car interior, etc.
- *Features:* Targets include topological and artistic particulars following the targeted vehicle interior theme such as traditional, modern, or futuristic.

- *Continuity:* Targets include shape continuity throughout vehicle interior, dash, and cargo spaces.

For interior size, three criteria are selected as examples:

- *Head Room:* Targets for this criterion include driver and passengers' head and leg room dimensions.
- *Leg Room:* Targets for this criterion include driver and passengers' leg room dimensions.
- *Entry and Egress:* Targets for this criterion include vehicle entry and egress dimensions and angles.

For interior surfaces materials, three criteria are selected as examples:

- *Textures:* Targets include interior materials' texture, finish, and surface properties.
- *Treatments:* Targets include interior materials, stitching, trimming, and treatment.
- *Colors:* Targets include interior colors, tones, patina, etc.

Due to its artistic nature, it may be easier to express most comfort targets graphically, using virtual reality, physically using mockups, or using a combination such as augmented reality. Accordingly, these subjective targets could be expressed and assessed using specified levels of approval based on experts and customers' evaluations.

4.3.1.3 Harmony: Encompasses full vehicle synchronization and coordination. Some of the interior targets may be expressed comparatively or graphically. As a vehicle balancing factor harmony significantly contributes to vehicle exterior and interior attractiveness and aesthetic. As shown in Figure 4.6, three sets of attributes are selected as examples.

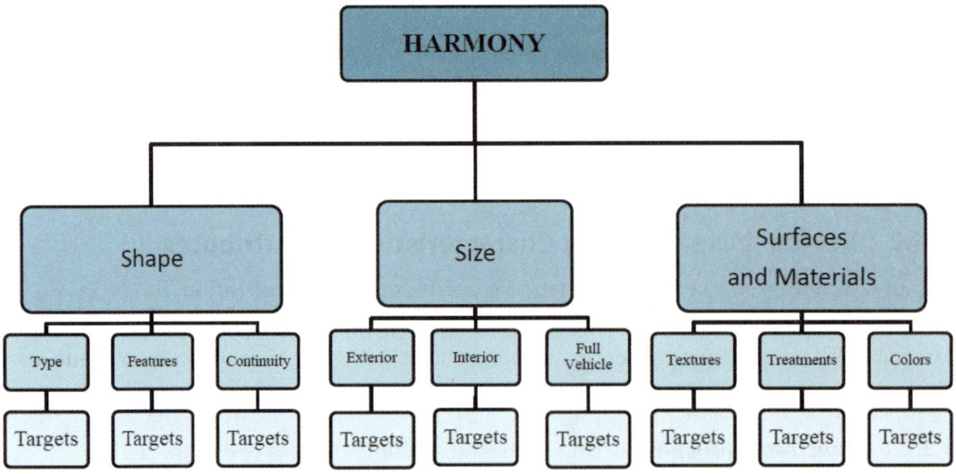

Figure 4.6 Selected attributes and criteria based on harmony.

- *Shape:* Affirm shape harmony of the vehicle exterior and interior in addition to the harmony between both
- *Size:* Affirm size proportionality of the vehicle exterior and interior in addition to the size proportionality between both
- *Surfaces and Materials:* Affirm surface and materials compatibility of the vehicle exterior and interior in addition to the compatibility between both

For shape, three criteria are selected as examples:

- *Type:* Targets include the vehicle exterior and interior harmony with vehicle type.
- *Features:* Targets include the harmony among different features of vehicle exterior and interior and between both.
- *Continuity:* Targets include continuity of shape harmony throughout the vehicle.

For size, three criteria are selected as examples:

- *Exterior:* Targets include size proportionality of exterior components.
- *Interior:* Targets include size proportionality of interior components.
- *Full Vehicle:* Targets include size proportionality between exterior and interior components.

For surfaces and materials, three criteria are selected as examples:

- *Textures:* Targets include texture compatibility of the vehicle exterior and interior.
- *Treatments:* Targets include compatibility of the vehicle exterior and interior treatments, such as paint and polish.
- *Colors:* Targets include compatibility of the vehicle exterior and interior colors in addition to the compatibility between both.

Due to its holistic and artistic nature, it may be easier to express most harmony targets graphically using artistic renderings, virtually using virtual reality, physically using mockups, or using a combination such as augmented reality. Accordingly, these subjective targets could be expressed and assessed using specified levels of approval based on evaluations by customers and experts.

4.3.2 Vehicle Manufacturing Characteristics and Attributes

Manufacturability is a key characteristic for product physical realization. In the following, three generic manufacturing characteristics and attributes are selected as examples. Three attributes for each characteristic and three criteria for each attribute are also selected as examples.

4.3.2.1 Construction: Constructing parts and subassemblies is one of the main manufacturing activities affecting product characteristics and attributes. The main attributes and most relevant criteria of vehicle construction characteristic are those affecting customer usage and physical realization of product architectural characteristics. Focusing on these aspects of construction, the following three attributes, shown in Figure 4.7, can be selected:

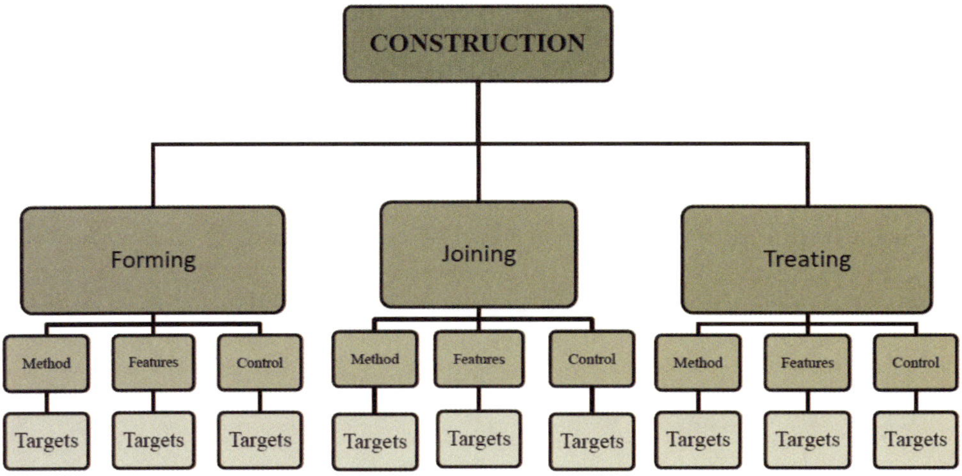

Figure 4.7 Selected attributes and criteria based on processing.

- *Forming:* Using manufacturing processes or processes to make the part or components
- *Joining:* Of shaped parts or components
- *Treating:* Materials and surfaces of parts or components [4-6]

For forming, the following three criteria are selected as examples:

- *Method:* Targets for this criterion could be stamping, forging, hydroforming, injection molding, etc.
- *Features:* Key or special features such as materials, geometry, and sequence.
- *Control:* Product parameters to be controlled during forming.

For joining, the following three criteria can be selected as examples:

- *Method:* Targets for this criterion could be spot welds, adhesives, screws, etc.
- *Features:* Specific features, sequence, or procedures during joining.
- *Control:* Product parameters to be controlled during joining.

For treating, the following three sets of criteria are selected as examples:

- *Method:* Targets for this criterion could be adding, removing, altering, and reshaping.

- *Features:* Targets for this criterion could be case hardening, grinding, shot peening, etc.
- *Control:* Product parameters to be controlled during treating.

Due to the vast number of materials and manufacturing processes, there are many options for forming, shaping, and constructing parts. Focusing on product characteristics and attributes in addition to other considerations, such as existing facilities, can guide the production process selection, design, development, and control.

4.3.2.2 Assembly: Product assembly characteristics greatly affect full vehicle form, function, and quality characteristics. Focusing on the main aspects of product assembly, the following three attributes, shown in Figure 4.8, can be selected:

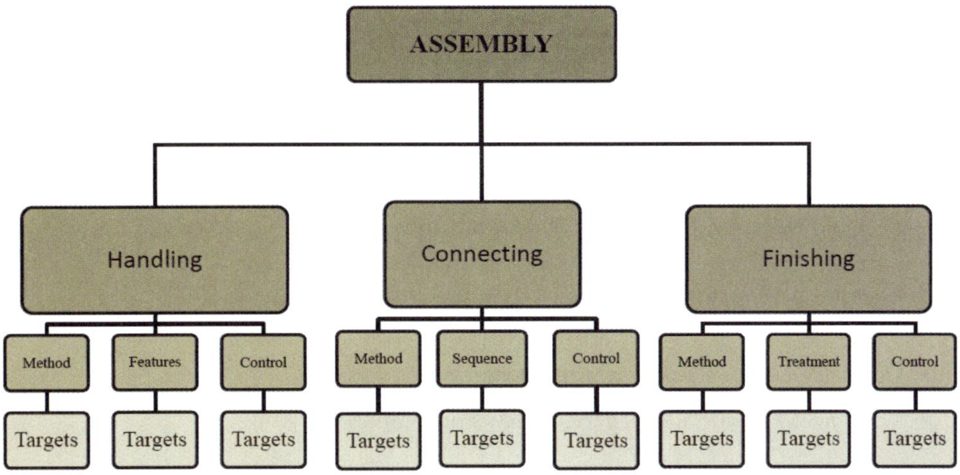

Figure 4.8 Selected attributes and criteria based on assembling.

- *Handling:* To convey parts and subsystems for full vehicle assembly
- *Connecting:* By bringing together parts and components to form subsystems and full vehicle
- *Finishing:* For all parts, subassemblies, and full vehicle

For handling, the following three criteria are selected as examples:

- *Method:* Targets for this criterion could be conveyors, gravity, fork lifts, etc.
- *Features:* Special handling features such as speed, platforms, clamps, and padding.
- *Control:* Product parameters to be controlled during handling.

For connecting, the following three criteria can be selected as examples:

- *Method:* Targets for this criterion could be couplings, press fit, welds, bolts and nuts, etc.
- *Features:* Specific features, sequence, or procedures during connecting.
- *Control:* Product parameters to be controlled during connecting.

For finishing, the following three sets of criteria are selected as examples:

- *Method:* Targets for this criterion could be painting, galvanizing, electroplating, etc.
- *Features:* Specific features, sequence, or procedures such as pretreatment, dipping, and rinsing.
- *Control:* Product parameters to be controlled during finishing.

Vehicle assembly is the peak of physical realization. Therefore, including the key full vehicle assembly attributes and characteristics during inception is necessary for the design and development of the product and production process as well. Subsystems and components-level assembly targets can be specified at the conception phase of the realization process.

4.3.2.3 Delivery: Vehicle delivery is the final phase of production. Delivery attributes such as inspection, packaging, and transporting affect the customer's received vehicle characteristics and attributes. Focusing on the main aspects of vehicle delivery characteristic, the following three attributes, shown in Figure 4.9, are selected:

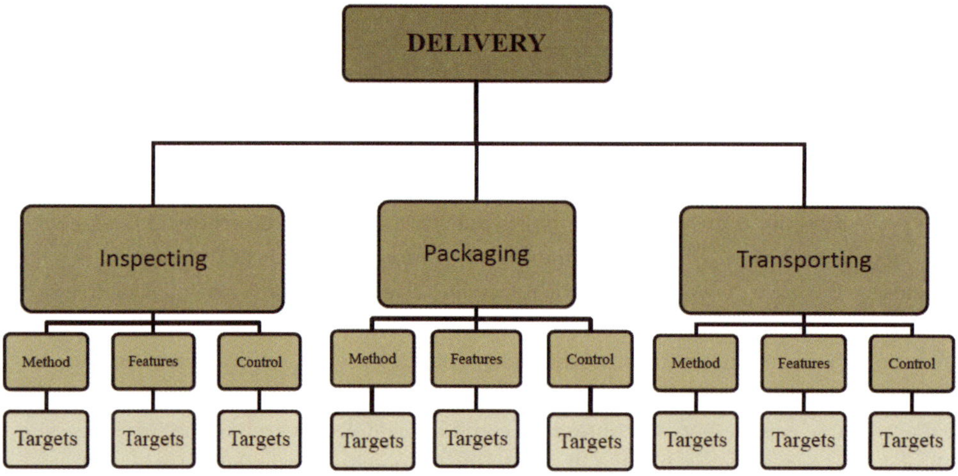

Figure 4.9 Selected attributes and criteria based on delivering.

- *Inspecting:* Of parts, components, subassemblies, and full vehicle
- *Packaging:* Of parts, components, subassemblies, and full vehicle
- *Transporting:* Of parts, components, subassemblies, and full vehicle

For inspecting, the following three criteria are selected as examples:

- *Method:* Targets for this criterion could be manual or automated.
- *Features:* Specific sequence or procedures during scanning such as visual inspection with memory function and laser scanning.
- *Control:* Product parameters to be controlled during inspecting.

For packaging, the following three criteria can be selected as examples:

- *Method:* Targets for this criterion could be manual or automated.
- *Features:* Specific procedures during packaging such as locations, orientation, and materials.
- *Control:* Product parameters to be controlled during packaging.

For transporting, the following three sets of criteria are selected as examples:

- *Method:* Targets for this criterion could be land, sea, air, etc.
- *Features:* Special considerations, sequence, or procedures during transporting.
- *Control:* Product parameters to be controlled during transporting.

Vehicle delivery from inspection to transporting is the last phase of physical realization before customers' test driving and/or purchasing. Therefore, considering full-vehicle delivery attributes during vehicle inception is desirable. Supplied parts, components, and subsystems delivery targets can be specified during the conception phase of the realization process. As indicated from the discussed examples of manufacturing characteristics, controlling product parameters during construction, assembly, and delivery is essential for achieving desired product attributes, characteristics, and quality.

4.4 Vehicle Function Characteristics and Attributes

The set of function characteristics and attributes are related to the product health and utility performances. Different products have different functions or performances by design or through evolution. In the following, as examples, three health and three utility characteristics are selected. The health characteristics are considered part of the function characteristics because of the total dependence of the utility characteristics and attributes on the health function for performance. While there are other relevant performance characteristics and attributes that could be included in each of the two categories, the selected examples are used to illustrate the framework presented in Figure 4.2.

4.4.1 Vehicle Health Characteristics and Attributes

4.4.1.1 Durability: The vehicle characteristic dealing with its ability to withstand the intended operating environment. As a key health characteristic, durability encompasses several vehicle resistance abilities or attributes. Focusing on mechanical loading and environmental exposure, the following three resistance areas, shown in Figure 4.10, are selected:

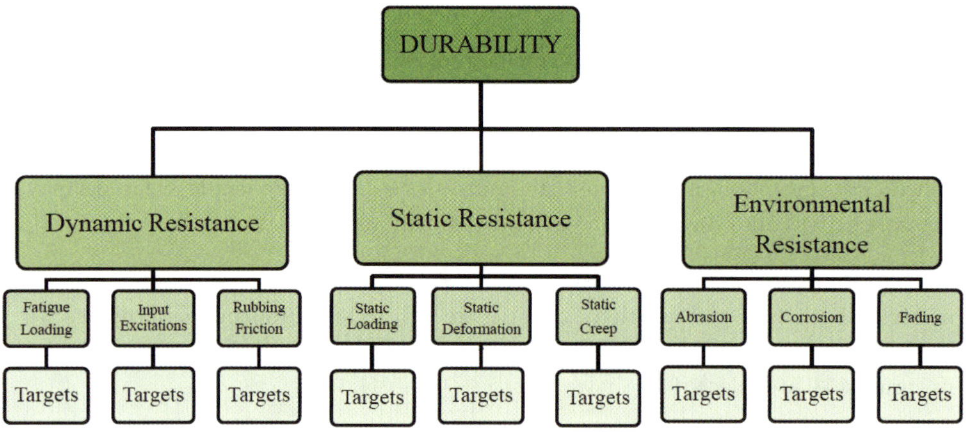

Figure 4.10 Selected attributes and criteria based on durability.

- *Dynamic Resistance:* Embodies the vehicle ability to resist dynamic loading and excitations
- *Static Resistance:* Embodies the vehicle ability to resist static loading and deflections
- *Environmental Resistance:* Embodies the vehicle ability to resist environmental damaging exposures

For dynamic resistance, three criteria are selected:

- *Fatigue Loading:* Targets include vehicle service life based on which the fatigue life and strength for all subsystems and components are established.
- *Input Excitations:* Targets include vehicle natural frequencies.
- *Rubbing Friction:* Targets include wear resistance for key subsystems and components such as power train and bearings.

For static resistance, three criteria are selected:

- *Static Loading:* Targets include static strength under different loadings such as jacking and towing.
- *Static Deformations:* Targets include static stiffness and total deformations under different loadings such as jacking and towing.
- *Static Creep:* Targets include all necessary stress and creep controls for exterior, interior, etc.

For environmental resistance, three criteria are selected:

- *Abrasion:* Targets for this criterion include abrasion resistance for exterior, interior, and key subsystems.
- *Corrosion:* Targets for this criterion include corrosion resistance for exterior, underbody, and key subsystems.
- *Fading:* Targets for this criterion include solar fading resistance for exterior, interior, and decorative components.

While some of the key vehicle-level durability targets, such as target life, can be established at the inception phase, some of the subsystem- and component-level targets may not be specified until the later phases of the realization process.

4.4.1.2 Reliability: One of the most important and comprehensive characteristics involving both form and function. Focusing on both architectural and performance aspects during vehicle development and validation, the following three criteria, shown in Figure 4.11, are selected:

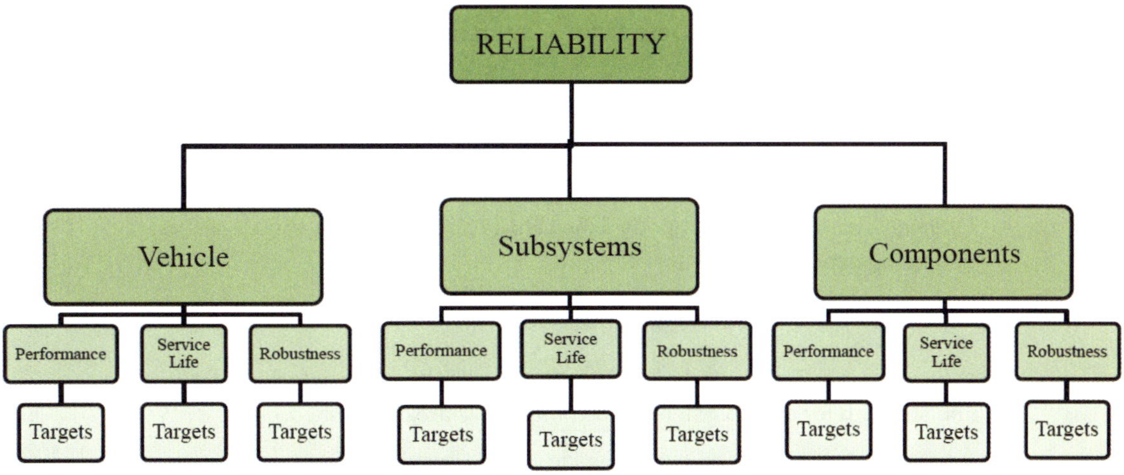

Figure 4.11 Selected attributes and criteria based on reliability.

- *Vehicle:* Embodies vehicle health and performance reliability
- *Subsystems:* Established health and performance reliability based on vehicle-level reliability
- *Components:* Established health and performance reliability based on subsystem-level reliability

For full vehicle, three criteria are selected:

- *Performance:* Targets include vehicle mobility, safety, connectivity, etc.
- *Service Life:* Targets include vehicle durability, amenities, interior, etc.

- *Robustness:* Targets include performance under harsh conditions and uncontrolled variations such as drivers, roads, and climate.

For subsystems, three criteria are selected:

- *Performance:* Targets include subsystem performance, stability, efficiency, etc.
- *Service Life:* Targets include subsystem durability, appearance, comfort, etc.
- *Robustness:* Targets include subsystem performance under harsh conditions and uncontrolled variations such as drivers, roads, and climate.

For components, three criteria are selected:

- *Performance:* Targets include components' functionality, performance, efficiency, etc.
- *Service Life:* Targets include components' durability, appearance, finish, etc.
- *Robustness:* Targets include components' performance under harsh conditions and uncontrolled variations such as drivers, roads, and climate.

While some of the key vehicle-level reliability targets should be established at the vehicle inception phase, some of the subsystem and component level targets could be specified during the conception and maturation phases.

4.4.1.3 Serviceability: Fueling, maintainability, and reparability are the key vehicle health attributes. Focusing on these attributes of vehicle serviceability characteristic, the criteria shown in Figure 4.12 are selected:

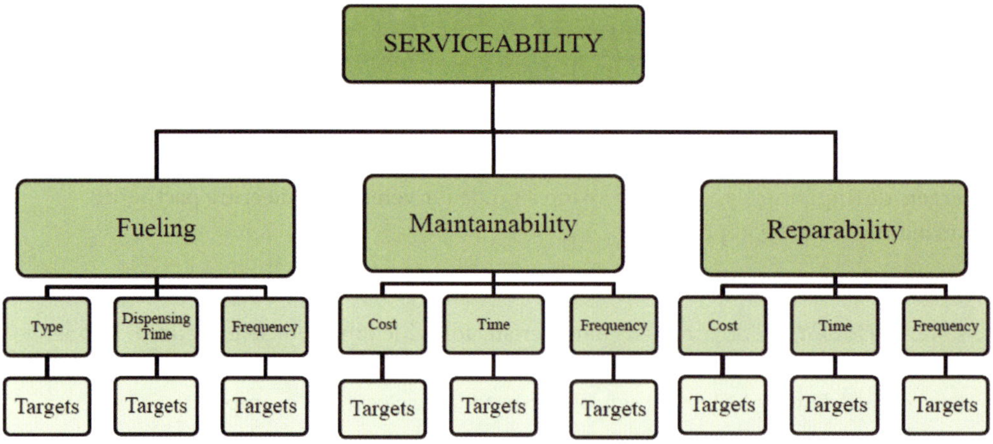

Figure 4.12 Selected attributes and criteria based on serviceability.

- *Fueling:* Embodies vehicle fueling services
- *Maintainability:* Embodies all vehicle maintenance services
- *Reparability:* Embodies all vehicle repair services

For fueling, three criteria are selected:

- *Type:* Targets include fuel type such as gasoline, diesel, natural gas, hydrogen, electric, etc.
- *Dispensing Time:* Targets include time to the fill fuel tank.
- *Frequency:* Targets include distance traveled before refueling, such as the number of traveled miles per full tank.

For maintainability, three criteria are selected:

- *Cost:* Targets include cost for periodic maintenance, tune up, part replacement, etc.
- *Time:* Targets include time needed for periodic maintenance, tune up, part replacement, etc.
- *Frequency:* Targets include the frequency of minor maintenance, such as changing oil and filters, in addition to major maintenance such as replacing battery, tires, and brake pads.

For reparability, three criteria are selected:

- *Cost:* Targets include cost for repairs, replacing broken components, etc.
- *Time:* Targets include time needed for repairs, replacing broken components, etc.
- *Frequency:* Targets include the frequency of required repairs due to nonaccidental failures.

While some of the key vehicle-level serviceability targets should be established at the vehicle inception phase, some cost, time, and frequency targets could be specified during the conception and maturation phases. However, establishing vehicle-level serviceability targets during inception would provide a guide for vehicle architecture packaging during the conception phase.

4.4.2 Vehicle Utility Characteristics and Attributes

4.4.2.1 Mobility: The function characteristic for which the vehicle is realized. As the key function of a vehicle, mobility determines vehicle road performance and capability. As shown in Figure 4.13, three sets of attributes are selected as examples.

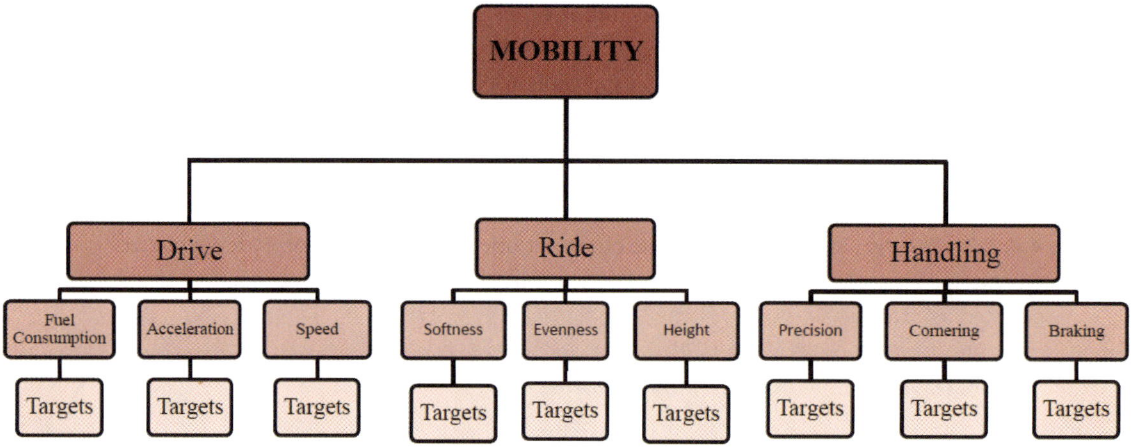

Figure 4.13 Selected attributes and criteria based on mobility.

- *Drive Attributes:* Embodies vehicle road performance capabilities
- *Ride Attributes:* Embodies vehicle ride comfort capabilities
- *Handling Attributes:* Embodies vehicle handling capabilities

For drive, three criteria are selected as examples:

- *Fuel Consumption:* Targets include fuel economy, fuel type, and range.
- *Acceleration:* Targets for this criterion include 0–60 time, ¼-mi acceleration, etc.
- *Speed:* Targets for this criterion include top speed, number of transmission speeds, etc.

For ride, three criteria are selected as examples:

- *Softness:* Appropriate with vehicle type. Targets include natural frequency of suspension system.
- *Evenness:* By reducing road irregularity effect. Targets include sprung to unsprung weight ratio.
- *Height:* Targets include height of the body at full load.

For handling, three criteria are selected as examples:

- *Precision:* Targets include frame torsional rigidity.
- *Cornering:* Targets include tire and rim size.
- *Braking:* Targets include braking distance.

Vehicle mobility attributes and targets must be defined at early development phases to fit vehicle type, class, and market segment. As the key characteristic of a vehicle, mobility targets determine vehicle driving capability and performance. In addition to the selection of key subsystems such as powertrain, suspension, and steering, these targets influence many features of the full vehicle such as packaging, vehicle mass, and mass distribution.

4.4.2.2 Safety: The essential vehicle characteristic. As a key function, safety governs customer's protection and wellbeing. As shown in Figure 4.14, three sets of attributes are selected as examples.

Figure 4.14 Selected attributes and criteria based on safety.

- *Crushability:* Portrays the vehicle crashworthiness ability
- *Protection:* Embodies the vehicle occupant protection ability
- *Avoidance:* Expresses the vehicle crash avoidance ability

For crushability, three criteria are selected as examples:

- *Frontal Impact:* Targets include specified frontal crushability targets for the vehicle's chosen star rating.
- *Side Impact:* Targets include specified side crushability targets for the vehicle's chosen star rating.
- *Rear Impact:* Targets include specified rear crushability targets for the vehicle's chosen star rating.

For protection, three criteria are selected as examples:

- *Frontal Impact:* Targets include occupant protection targets for the vehicle's chosen frontal impact star rating.
- *Side Impact:* Targets include occupant protection targets for the vehicle's side impact chosen star rating.
- *Rear Impact:* Targets include occupant protection targets for the vehicle's chosen rear impact star rating.

For avoidance, three criteria are selected as examples:

- *Warning:* Targets include parameters needed for the vehicle's chosen crash warning strategies.
- *Mitigation:* Targets include parameters needed for the vehicle's chosen crash mitigation strategies.
- *Prevention:* Targets include parameters needed for the vehicle's chosen crash prevention strategies.

Desired safety ratings are usually established at the early vehicle realization phases. Consequently, vehicle-level targets are specified following the applicable safety standards such as FMVSS [4-7], [4-8], [4-9]. As a top characteristic of a vehicle, safety targets influence many features of the vehicle such as exterior, interior, structure, and packaging.

4.4.2.3 Amenities: An important vehicle characteristic, for daily interaction, from the customer's perspective. As a key functioning characteristic, amenities encompass the human factors that facilitate the interface with the vehicle and enrich customers' daily experience. As shown in Figure 4.15, three sets of attributes are selected as examples:

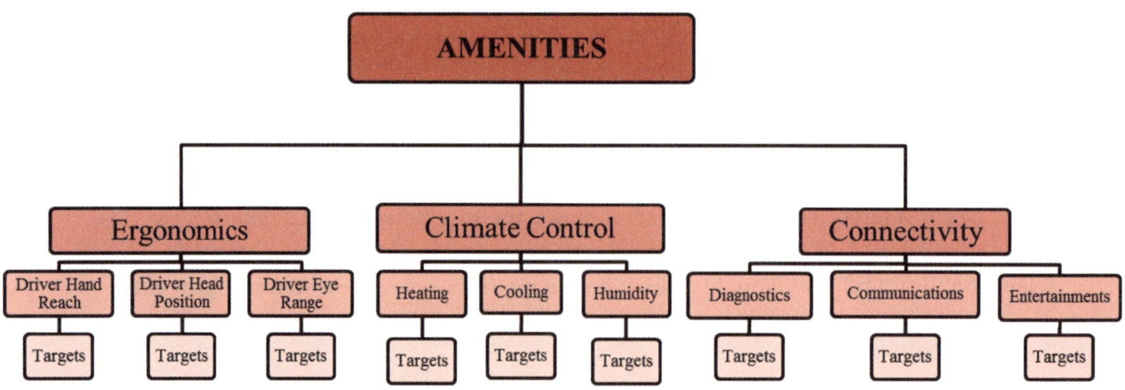

Figure 4.15 Selected attributes and criteria based on amenities.

- *Ergonomics:* Embodies the vehicle interface and accommodation abilities
- *Climate Control:* Embodies vehicle climate adjustability and suitability
- *Connectivity:* Embodies vehicle internal and external communication and accessibility

For ergonomics, three criteria are selected:

- *Driver's Hand Reach:* Targets can be established according to the SAE J287 standard [4-10].
- *Driver's Head Position:* Targets can be established according to the SAE J1052 standard [4-10].
- *Driver's Eye Range:* Targets can be established according to the SAE J941 standard [4-10].

For climate control, three criteria are selected:

- *Heating:* Targets include heating capacity, temperature range, etc.
- *Cooling:* Targets include cooling capacity, cooling rate, air circulation, etc.
- *Humidity:* Targets include humidity control parameters and sensors accuracy.

For connectivity, three criteria are selected:

- *Diagnostics:* Targets include vehicle systems diagnostics accuracy, redundancy, control parameters, and ranges, etc.
- *Communication:* Targets include bandwidth, fault tolerance, flexibility, etc.
- *Entertainment:* Targets include audio amplification levels and visual resolutions.

To achieve vehicle interior balance, targets for the human factor aspect of amenities and architectural harmony must be concurrently established and accomplished. Therefore, it may be logical to have harmony and human factors under the same vehicle integration area, as will be discussed in the later chapters.

4.5 Vehicle Fit Characteristics and Attributes

Along with the form and function, the key set of product fit characteristics and attributes must be considered. The product fit set contains the summative product characteristics and attributes. Examples of the most significant fit characteristics are affordability, dependability, and initial quality illustrated in the framework presented in Figure 4.16.

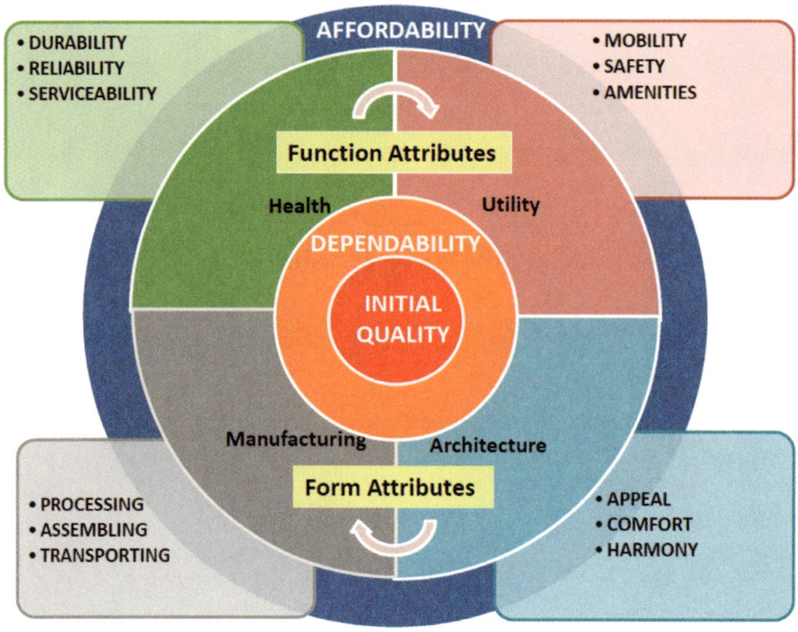

Figure 4.16 Framework for vehicle form, function, and fit characteristics.

As shown in Figure 4.16, the fit characteristics' affordability, dependability, and initial quality are summative. To achieve vehicle affordability, each form and function characteristic and attribute must meet established affordability targets. Similarly, for initial quality and dependability, each form and function characteristic and attribute must meet customers' expectations. In the following, some illustrative examples for affordability, dependability, and initial quality will be discussed. A more detailed discussion on quality will be conducted in Chapter 5.

4.5.1 Affordability

Product affordability is a fit characteristic associated with all vehicle form and function characteristics and attributes. Affordability targets must be established at vehicle inception based on intended customers and targeted markets. As shown in Figure 4.17, three sets of affordability costs are selected:

- *Initial Cost:* With targets for development, production, and retail per unit cost
- *Health Cost:* With targets for replacements, repair, and maintenance lifecycle upkeep cost
- *Utility Cost:* With targets for fuel, lubricants, and connectivity cost

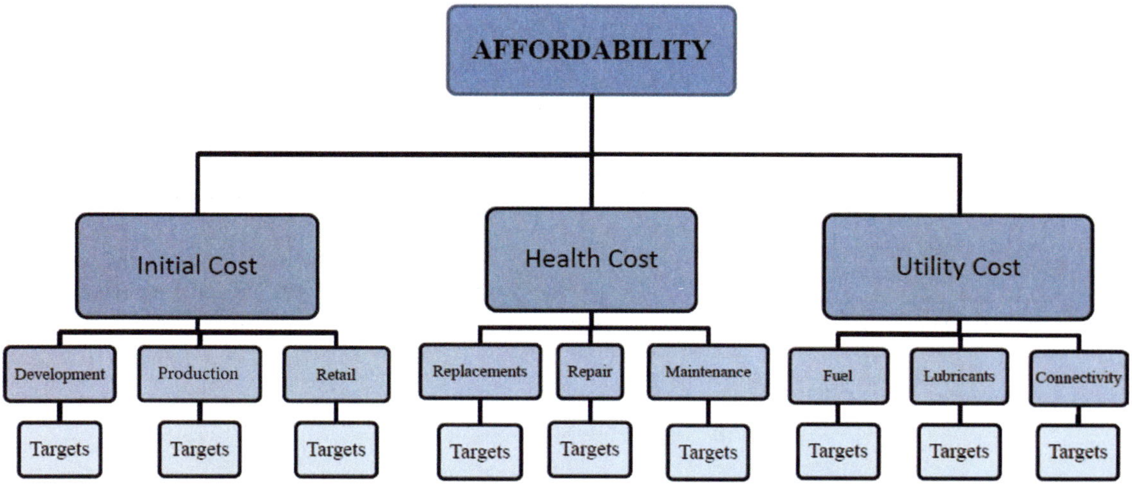

Figure 4.17 Selected attributes, and criteria based on affordability.

Initial cost is the main contributor to vehicle ownership total cost. Repair, periodic maintenance, and components' replacement due to wear and tear are also major contributors to ownership total cost over vehicle lifecycle. Similarly, fuel, lubricants, and any connectivity costs are additional contributors to vehicle ownership total cost over vehicle lifecycle.

4.5.2 Dependability

Product dependability is a summative fit characteristic that reflects customers' experience and confidence in product form and function quality, reliability, and durability over a period of usage. In other words, extended product usage transforms the continuous delivery of built-in form and function quality, reliability, and durability to the customers' perceived dependability. Accordingly, with higher quality, reliability, durability, and lower probability of failure, higher customers' perceived dependability ratings could be achieved. For example, testing the number of samples associated with targeted reliability and confidence, during product validation, is one of the key factors for achieving reliability and consequent dependability targeted ratings. For illustration, three sets of dependability characteristics with previously discussed attributes and criteria are shown in Figure 4.18:

- *Overtime Utility:* Extended usage mobility, safety, and amenities characteristics
- *Overtime Health:* Extended usage durability, reliability, and serviceability characteristics
- *Overtime Form:* Extended usage shape, materials, and finish characteristics

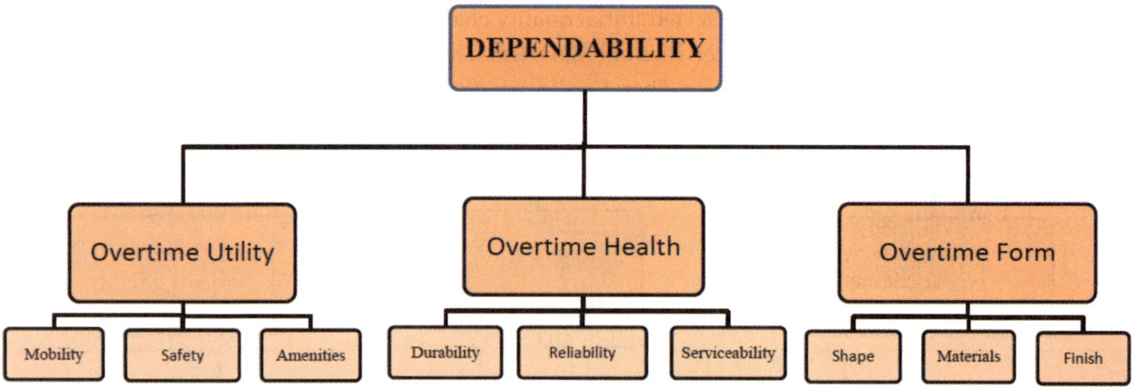

Figure 4.18 Selected dependability characteristics.

The targets for the selected dependability characteristics of Figure 4.18 are for extended usage or overtime performance such as 3, 5, or 10 years of customer's usage. It could also be established for extended performance over 30, 50, or 100 thousand of customer's usage miles.

4.5.3 Initial Quality

Initial quality is the peak summative characteristic that reflects meeting customers' expectations in all built-in product form and function characteristics. Accordingly, in product realization, "quality is everyone's responsibility," as succinctly stated by W. Edwards Deming. For illustration, three sets of initial quality characteristics with previously discussed attributes and criteria are shown in Figure 4.19:

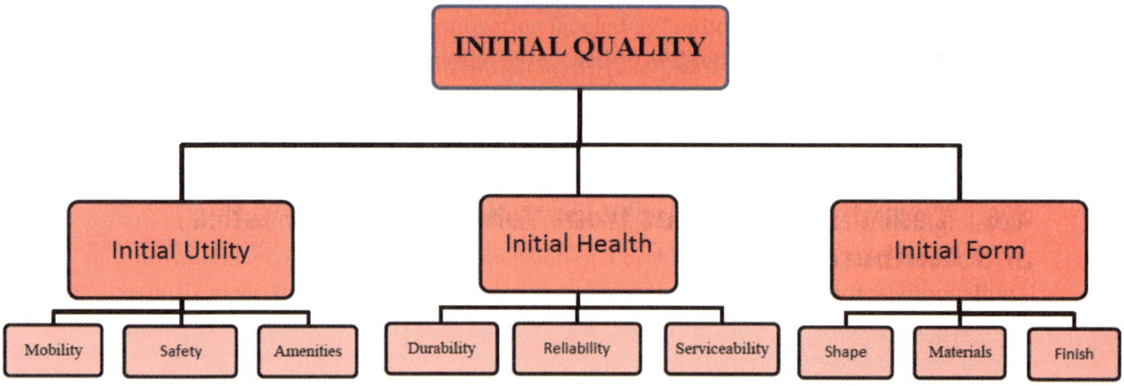

Figure 4.19 Selected initial quality characteristics.

- *Initial Utility:* As produced mobility, safety, and amenities characteristics
- *Initial Health:* As produced durability, reliability, and serviceability characteristics
- *Initial Form:* As produced shape, materials, and finish characteristics

As shown in Figure 4.19, the selected initial quality characteristics are the same as the selected dependability characteristics of Figure 4.18. The differences are mainly in the initial and overtime performance. Therefore, initial quality and dependability targeted performances can be combined under one fit characteristic, as shown in Figure 4.20.

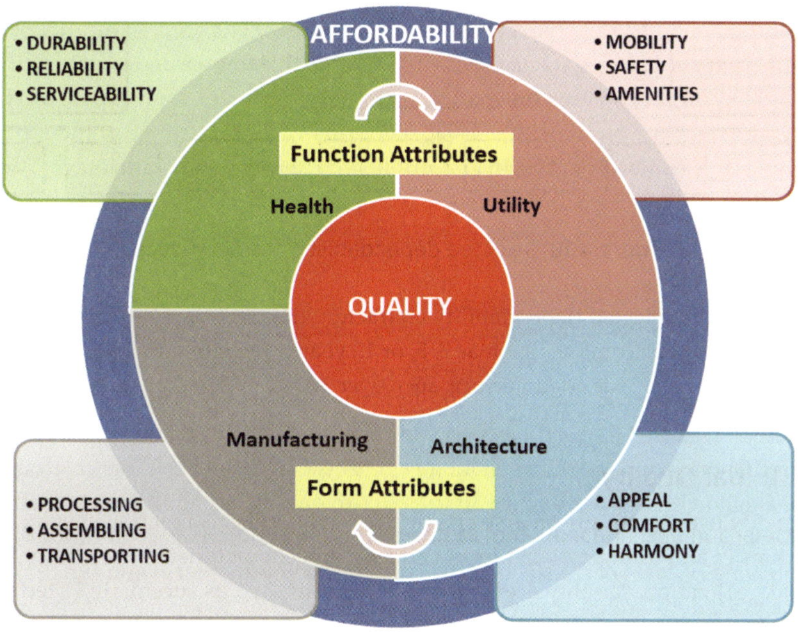

Figure 4.20 Framework for vehicle form, function, affordability, and quality characteristics.

During product inception, conception, and maturation, established quality targets and development efforts should include both initial and enduring performances for all form and function characteristics. More detailed discussions for initial and long-term quality will be conducted in Chapter 5.

4.6 Learning Outcomes from Vehicle Characteristics and Attributes

The desired learning outcomes from studying vehicle characteristics and attributes can be categorized as:

4.6.1 Knowledge

Achieving customers' desired characteristics and attributes, during new vehicle development, is the main driver in a competitive market. Accordingly, identifying and achieving these desired characteristics and attributes is the key to success for any realization process. Therefore, using a well-defined framework to develop the proper criteria and facilitate the development of customers' desired vehicle specifications is essential.

In general, several criteria constitute an attribute and several attributes constitute a product characteristic. For vehicles, product characteristics and attributes can be classified into two basic form and function sets in addition to the summative set of fit characteristics.

The basic set of form characteristics and attributes is composed of two subsets of architectural and manufacturing characteristics and attributes. The architectural subset contains the architectural characteristics such as appeal, comfort, and harmony. The manufacturing subset contains the manufacturability attributes such as construction, assembly, and delivery. Similarly, the basic set of function characteristics and attributes is composed of two subsets of health and utility characteristics and attributes. The health subset contains characteristics such as durability, reliability, and serviceability. The utility subset contains the key performance characteristics such as mobility, safety, and amenities.

The summative set of fit characteristics includes affordability, dependability, and initial quality. The key fit characteristics of affordability is associated with vehicle form and function characteristics and attributes through affordability of initial costs such as production cost, as well as running costs such as vehicle health and utility costs. The summative fit characteristic of dependability reflects customers' confidence, based on overtime extended usage experience, in the built-in quality, reliability, and durability of product form and function. As the peak summative global characteristic, initial quality reflects customers' level of satisfaction in as-produced product form and function characteristics. As the differences between the characteristics of dependability and initial quality are only in performance times, the two summative fit characteristics can be combined under the fit characteristic of quality. As a result, established quality targets and development efforts, during product realization, should include both initial and overtime performances for all form and function characteristics.

4.6.2 Skills

Building the skills to correctly identify product key characteristics and attributes, in addition to systematically pursuing them throughout the realization process, is essential. This also requires building the necessary skills to recognize the relevant attributes for each characteristic, the correct criteria for each attribute, and the right targets for each criterion. Additional skills are needed for the development of vehicle, subsystems, and CTS from the customers' desired characteristics and attributes.

4.6.3 Paradigm/Attitude

Vehicle, subsystems, and components technical specifications should be derived from the customers' desired characteristics and attributes. Technical specifications of previously developed vehicles can be used as guide or template with entries derived from established desired vehicle targets.

4.7 References

4-1. El-Sayed, M., "Lean Design for Integrated Product Realization," *SAE Int. J. Mater. Manuf.* 3(1):194-201, 2010, doi:10.4271/2010-01-0400.

4-2. El-Sayed, M. and El-Sayed, J., "Balancing Manufacturability and Performance Attributes in Lean Design," *SAE Int. J. Mater. Manf.* 5(1):174-182, 2012, doi:10.4271/2012-01-0512.

4-3. Dankwort, C. W., and Podehl G. 2000. A New Aesthetic Design Workflow—Results from the European Project FIORES, CAD Tools and Algorithms for Product Design (Dagstuhl Seminar, November 1998), pp. 16–30, Springer-Verlag: London, UK.

4-4. ISO. International Standards ISO 3833. 1977. Road Vehicles—Types—Terms and definitions, second Edition. ISO, Available from: http://www.iso.org/iso/catalogue_detail.htm?csnumber=9389.

4-5. Rossbacher, U. 2007. Advanced Modeling and Surfacing Techniques with Autodesk AliasStudio (Part I), Autodesk University: Vegas.

4-6. Vetter, J., Barbezat, G., Crummenauer, J., and Avissar, J. "Surface treatment selections for automotive applications," *Proceedings of the 32nd International Conference on Metallurgical Coatings and Thin Films*, 200(5–6):1962–1968, Elsevier, November 2005.

4-7. Hollowell, W., Gabler, H., Stucki, S., Summers, S., and Hackney, J., *Updated Review of Potential Test Procedures for FMVSS no. 208*, Office of Vehicle Safety Research, October 1999.

4-8. Kahane, C. "Evaluation of FMVSS 214 Side Impact Protection Dynamic Performance Requirement." NHTSA Technical Report, DOT HS 809 004, October 1999.

4-9. Department of Transportation, National Highway Traffic Safety Administration. 2007. "Laboratory Test Procedure for FMVSS 301R, Fuel System Integrity—Rear Impact," TP-301R-02. Department of Transportation: Washington, DC.

4-10. http://www.sae.org/standards.

Chapter 5
Product Quality and Process Improvability

"Quality comes not from inspection but from improvement of the process."

—Mary Walton

5.1 Introduction

The results of a customer survey on product quality [5-1], [5-2] revealed that the top three characteristics that determine product quality are "works as it should" (98%), "lasts a long time" (95%), and "easy to maintain" (93%). The same survey shows that of less importance are "looks attractive" (58%), "incorporates the latest technology" (57%), and "has many features" (48%). Accordingly, these results show that functionality, dependability, and maintainability are more valued characteristics by the customer than attractiveness, latest technology, and having many features. Based on these results, it is clear that product characteristics and attributes are not equally desired or valued by the customers. Therefore, understanding quality and its relationship to customers' desired product characteristics and attributes is necessary for the realization of a product that meets or exceeds customers' expectations.

In this chapter, some in depth analyses will be conducted for quality and its relationship to the development gaps in product form and function characteristics and attributes during different product realization phases. Subsequently, potential losses due to these quality gaps will be examined. Finally, process manageability and improvability for achieving product quality targets during product realization will be discussed along with some examples of quality tools and methodologies.

5.2 Product Quality Gaps

As discussed in Chapter 4, one of the key factors that differentiate between initial quality and dependability is time. Therefore, targeted performances for both initial quality and dependability can be combined under the fit characteristic of quality, as shown in Figure 5.1.

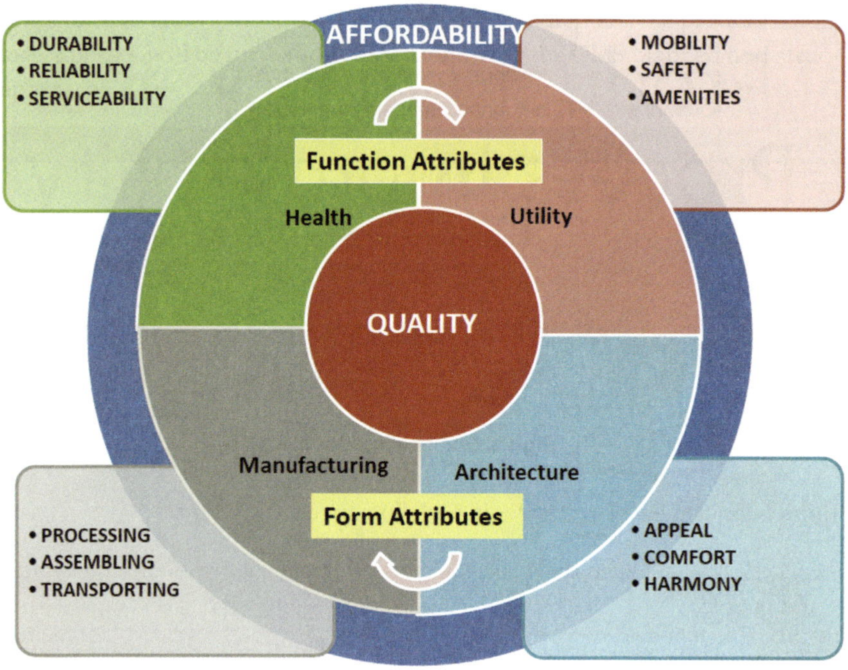

Figure 5.1 Quality as joint product dependability and initial quality fit characteristic.

By combining initial quality and dependability, established quality targets and development efforts should include both initial and long-term aspects of all form and function characteristics. In this regard, the ISO 8402-1986 standard [5-3] defines quality as "the totality of features and characteristics of a product or service that bears its ability to satisfy stated or implied needs." In light of the realization concepts discussed in Chapter 2, in addition to the product form, function, and fit characteristics discussed in Chapter 4 and illustrated in Figure 5.1, further specification of quality can be made by the following definition:

Quality is a fit characteristic that sums the perceptual level of satisfaction relative to expected or desired form and function characteristics and attributes.

To understand product quality, three distinct domains with three basic elements in each domain must be considered, as shown in Figure 5.2.

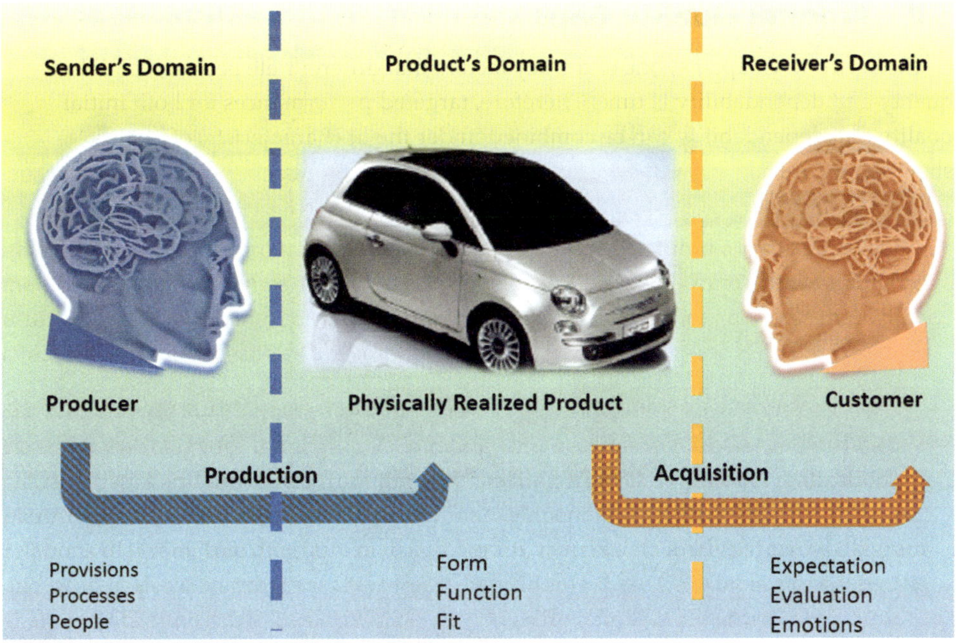

Figure 5.2 Quality domains.

The three basic elements to consider in each domain are as follows:

- *Sender's Domain:* With the three enterprise elements of provisions, processes, and people
- *Product's Domain:* With the three sets of form, function, and fit characteristics
- *Receiver's Domain:* With the three perceptual functions of Expectation (imaginative), Evaluation (cognitive), and Emotions (affective)

While all product realization activities are performed in the sender's domain and the intrinsic or built in characteristics are in the product's domain, quality is always assessed in the receiver's domain. For example, during initial product evaluation, the differences between customer's expectation and customer's cognitive evaluation of a perceived product attribute is translated into customer's perceived cognitive gap. Consequently, the perceived cognitive gap is translated into a perceived quality gap in the customer's perceptual domain due to customer's affective reaction to it. Therefore, with differences between individuals in expectations, evaluations, and preferences, assessment of quality in the receiver's perceptual domain is subjective. However, despite the subjective nature of quality assessment, ample work can be done in the sender's domain to continuously improve the intrinsic characteristics of the product and close all possible perceived cognitive gaps.

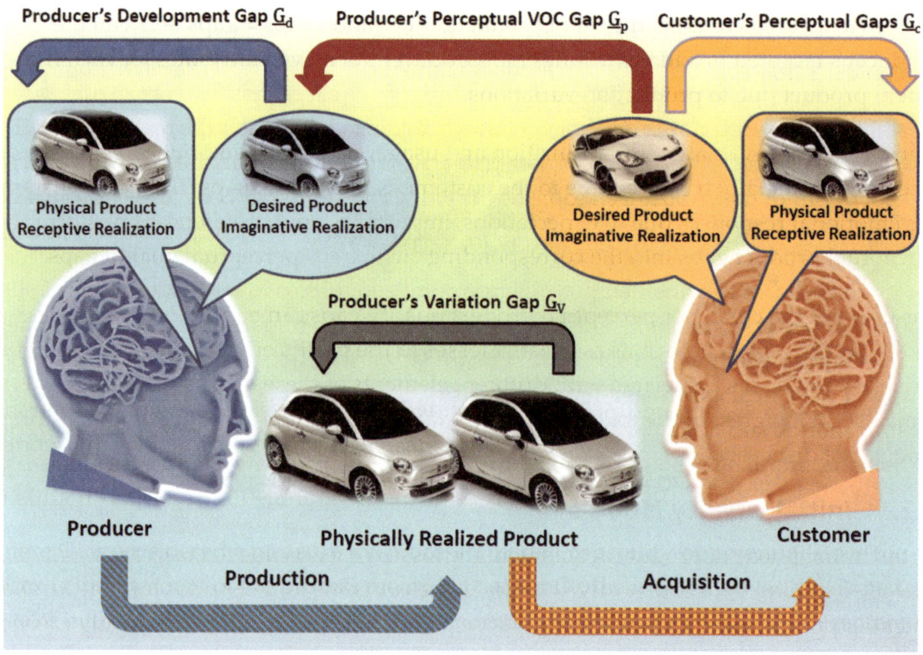

Figure 5.3 Product gaps.

In light of the realization concepts discussed in Chapter 2, customer's perceived gaps in product attributes can be depicted, as shown in Figure 5.3, in which:

$$G_c = G_p + G_d + G_v \tag{5.1}$$

where G_c is an array that contains the set of customer's perceptual gaps = gaps between the customer's expectation (imaginative perceptual realization) and cognitive evaluation (receptive perceptual realization) of the physically realized product attributes.

G_p is an array that contains the set of producer's perceptual voice of the customer gaps = gaps between the producer's imaginative perceptual realization and customer's imaginative realization of expected and desired product attributes. This set could have negative gaps due to missing customer's expected and desired attributes or the presence of unexpected and undesired attributes. It may also contain positive gaps due to the presence of customers' unexpected but desired attributes. Examples of these gaps include most observable architectural attributes such as style, color, interior, seating, and comfort.

G_d is an array that contains the set of producer's development gaps = gaps between the producer's imaginative perceptual realization and producer's physically realized product attributes. This set may contain negative gaps due to unachieved targeted attributes. It could also contain positive gaps due to exceeding targeted attributes. Accordingly, gaps in this set are mainly caused by the achievement gaps in product targets. From the customer's perspective, however, the attributes associated with these gaps are usually recognizable. Examples of these gaps include most known functional attributes and targets such as fuel economy, acceleration, ride, handling for mobility, and star ratings for safety.

G_v is an array that contains the set of production variation gaps = gaps between the producer's targeted product attributes and producer's achieved attributes for each individual product due to production variations.

During product evaluation for acquisition and usage, customer's emotional responses or reactions are usually triggered due to the customer's perceptual gaps (G_c). These customer's emotional responses due to expectations, importance, and preferences transform the perceptual product gaps into the corresponding customer's perceptual quality gaps.

The effect of the customer perceptual product quality gaps can be studied by focusing on the tangible aspects such as associated losses to the producer, customers, and society. Modeling the losses associated with different elements of these quality gaps can yield some useful insights. In the following, some examples of initial and long-term quality models are discussed.

5.2.1 Initial Quality Models

At the initial phase of product acquisition, the losses or gains due to customer's perceived quality gaps are mostly potential losses or gains to the producer in the form of potential profits, sales, and customers. These potential losses or gains are mainly caused by the producer's failure or success to provide what the customer wants and expects, as defined by the voice of the customer (VOC). The set of producer's VOC gaps (G_p) and some elements of the set of producer's development gaps (G_d) are usually the main sets of gaps associated with the initial product evaluation.

Elements of the set of the producer's VOC gaps (G_p) are sometimes observed simultaneously with some of the elements of the set of producer's development gaps (G_d). The surprising gaps in both sets can be considered as elements of the set (G_S). These elements are mostly from G_p with some elements from G_d. These surprising gaps may cause the customer reaction to be exponential. Assuming that the producer's potential loss or gain is proportional to the customer reaction, the potential loss or gain associated with any element of the surprising set of gaps G_S could be expressed as exponential function of that element magnitude. Therefore, the following exponential function could mathematically model or express the potential loss or gain associated with any element (G_{Si}) from the set of surprising gaps G_S:

$$L(G_{Si}) = K_{Si}e^{\|G_{Si}\|} \tag{5.2}$$

where L represents potential loss or gain and G_{Si} represents a negative or positive gap. K_{Si} is the coefficient associated with the element G_{Si}. The sign of the coefficient K_{Si} follows the sign of the gap, i.e., K_{Si} is negative if the gap is negative and K_{Si} is positive if the gap is positive. Accordingly, Eq. 5.2 represents an exponential relationship between producer's potential losses (negative L) or gain (positive L) due to any surprising negative or positive gap, respectively.

The nonsurprising elements of G_p and G_d can be considered as elements of the set (G_N). Due to prior anticipation, the customer's reaction to these elements could be proportional to the gaps. These nonsurprising elements of the set G_N are mostly from G_d with some elements from G_p. Considering that the producer's potential losses or gains are proportional to the customer's reaction, the potential producer's loss or gain associated with any G_N element is also proportional to that element. Therefore, a linear function could mathematically model or express the producer's potential losses or gains associated with any element (G_{Ni}) from the set of nonsurprising gaps G_N:

$$L(G_{Ni}) = K_{Ni}G_{Ni} \tag{5.3}$$

where L represents potential loss or gain and G_{Ni} represents a negative or positive gap. K_{Ni} is a positive coefficient associated with the element G_{Ni}. Since any of these gaps could be positive or negative, a negative gap would cause a potential loss while a positive gap would cause a potential gain. Therefore, Eq. 5.3 represents a linear relationship between producer's potential loss (negative L) or gain (positive L) and any nonsurprising gap of a desired and expected product attribute.

In Eq. 5.2 and Eq. 5.3, negative gaps could result in linear or exponential potential losses to the producer in the form of reduced profit, losing the transaction, or losing a customer. On the other hand, a positive gap due to exceeding customer's expectations or the presence of an unexpectedly delighting attribute could lead to linear or exponential potential gains in the form of increased profit, securing the transaction, and attracting a customer.

Sample of the plotted graphs of Eq. 5.2 and Eq. 5.3 are presented in Figure 5.4, with $||K_{Si}|| = K_{Ni} = 1$.

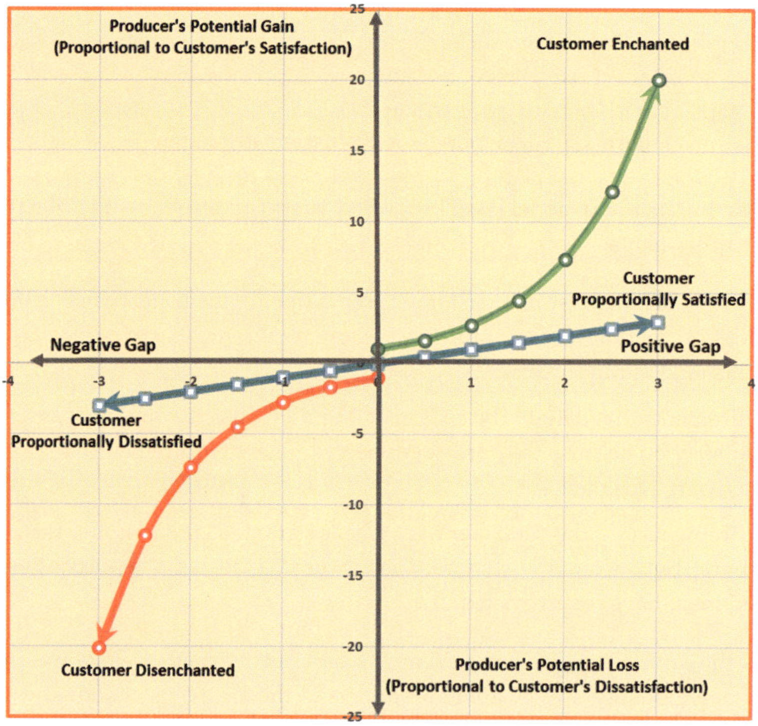

Figure 5.4 Producer's linear and exponential potential gains and losses.

These graphs show the potential producer's losses and gains due to different customer reactions caused by negative and positive gaps in in the surprising set G_S (exponential response) and the nonsurprising set G_N (linear response). These initial quality potential losses or gains are in agreement with the customers' responses observed by the well-known Kano model [5-4], [5-5], [5-6], as illustrated in Figure 5.5 [5-5].

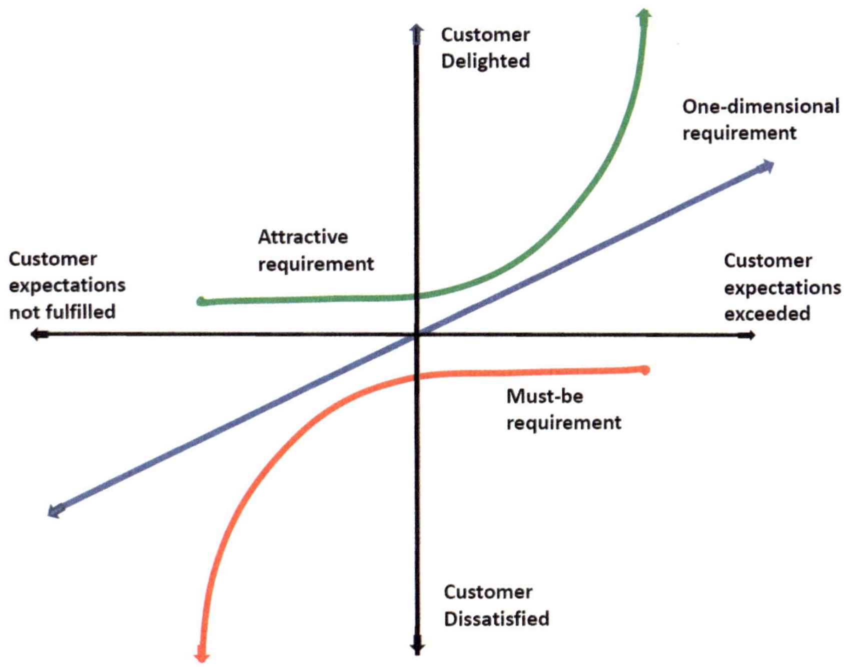

Figure 5.5 Kano model for customer satisfaction [5-5].

As identified by the Kano model [5-4], [5-5], [5-6], these are responses to three types of customer requirements:

1. Requirements that result in customer proportional satisfaction when fulfilled and dissatisfaction when not fulfilled. These requirements are called one-dimensional requirements.
2. Requirements that delight the customer, but their absence does not cause dissatisfaction because they are not expected. These requirements are called attractive requirements.
3. Requirements that the customer may take for granted (and assumes their presence), but becomes very dissatisfied if they are absent. These requirements are called must-have requirements.

To identify the different types of requirements, an effective methodology was developed. This methodology uses customers' surveys in the form of questionnaires that contain one positive and one negative question for each requirement being surveyed [5-4], [5-5], [5-6].

The producer's loss models introduced in Eq. 5.2 and Eq. 5.3 could be graphed for key attributes and targets using customer surveys, Kano questionnaires, and warranty data. However, by differentiating between the surprising and nonsurprising attributes in the set of producer's VOC gaps (G_p) and the set of the producer's development gaps (G_d), the potential associated losses or gains could be easily recognized as being exponential or linear. Recognizing a gap for being positive or negative and the type of reaction for being linear or exponential is necessary for guiding product planning and development.

Issues associated with elements of G_p and G_d are usually initial quality problems. These problems are mostly associated with G_p due to the absence of desired attributes or the existence of undesired attributes, and with G_d due to unachieved product targets. These issues can also appear during the initial period of the customer's usage after buying the vehicle. For example, J. D. Power Initial Quality Study (IQS) surveys owner-reported problems experienced per 100 vehicles (PP100) during the first 90 days of new-vehicle ownership [5-7]. Even after selling the vehicle, the quality issues can cause losses to the producer in the form of recalls or loss of future sales due to diminishing customers' loyalty, as illustrated by the following case study.

5.2.1.1 Case Study 5.1: The J. D. Power 2014 IQS finds that "the fewer problems owners experience with their vehicle, the greater their loyalty to the brand. Combined data from the previous years' IQS results and the power information network (PIN) from J.D. Power show that 57% of owners who reported no problems stayed with the same brand when they purchased their next new vehicle. Brand loyalty slips to 53% among owners who reported just a single problem and to only 48% percent among owners who reported two or more problems" [5-7]. The study also detects "increased problems among all-new vehicles in the areas of voice recognition, Bluetooth pairing and audio systems" due to "new features and technology" [5-7].

5.2.2 Long-Term Quality Models

In addition, to some unachieved targets associated with G_d, problems associated with elements of the set of production variation gaps G_v usually appear during long-term customer's usage. These elements of G_v and some elements of G_d can be considered as elements of the long-term set (G_L). To address the long-term issues associated with these elements, a quadratic loss function has been suggested and successfully utilized to improve product quality and reliability [5-8], [5-9]:

$$L(G_{\text{Li}}) = K_{\text{Li}} G_{\text{Li}}^2 \tag{5.4}$$

where L represents the loss and G_{Li} represents a negative or positive gap. And K_{Li} is the coefficient associated with the element G_{Li}. Following [5-9], if y represents the actual achievement of a targeted level m, then variation gap G_{Li} can be written as:

$$G_{\text{Li}} = y - m \tag{5.5}$$

Using Eq. 5.5, Eq. 5.4 can be written as

$$L(y) = K_{Li}(y - m)^2 \tag{5.6}$$

If the product specification is given as $m \pm \Delta_0$ and A_0 is the potential loss due to the cost associated with repair when the magnitude of the gap $\|y - m\| = \Delta_0$, then $K_{Li} = A_0/\Delta_0^2$ from Eq. 5.6.

Equation (5.5) and the potential associated loss equation Eq. 5.6 represent the conditions when the nominal or the targeted value m is the best achievement level. Conditions when the targeted value is zero and the smallest value is the best can be modeled using $G_{Li} = y$ [5-9].

Similarly, conditions when the largest value is the best can be modeled using $G_{Li} = 1/y$ [5-9].

Using the quadratic loss function and its derivatives has become a common engineering practice for addressing variations in many robust design applications [5-10].

Because production variations and some unachieved design targets issues should be addressed using the robust design tools and reliability validation testing, these long-term quality issues could be reliability issues. As discussed earlier, these issues from the customer's perspective are dependability problems. For example, J. D. Power Dependability Study surveys these owner-reported PP100 during the third year of vehicle ownership [5-11]. Similar to initial quality, long-term quality or dependability issues can cause losses to the producer in the form of losing future sales due to losing customers' loyalty, as shown by the following case study.

5.2.2.1 Case Study 5.2: The J. D. Power 2014 Dependability Study indicated that "the fewer problems owners experience with their vehicle, the greater their loyalty to the brand. Combined data from previous years' VDS results and vehicle trade-in data from the PIN from J. D. Power show that 56% of owners who reported no problems stayed with the same brand when they purchased their next new vehicle. Brand loyalty slipped to just 42% among owners who reported three or more problems" [5-11]. The study also detects increased problems in "engine hesitation, rough transmission shifts and lack of power" due to "continually looking for ways to improve fuel economy" [5-11].

5.3 Process Improvability

To attain desired process characteristics and attributes during inception, established process targets must be achieved and validated by the end of process maturation phase. However, for existing processes, maintaining or enhancing the original process characteristics and attributes by establishing and attaining new targets must be pursued through continuous improvement efforts. To assure that continuous improvement activities are organic to the process and systematically performed, process improvability could be joined with process manageability as discussed in Chapter 3. For production process, the resulting fit characteristic of quality control is the key for continuous improvement of product quality and achievement of quality targets during production. This could be observed by considering the fostering relationship between the product and production process as illustrated in Figure 5.6.

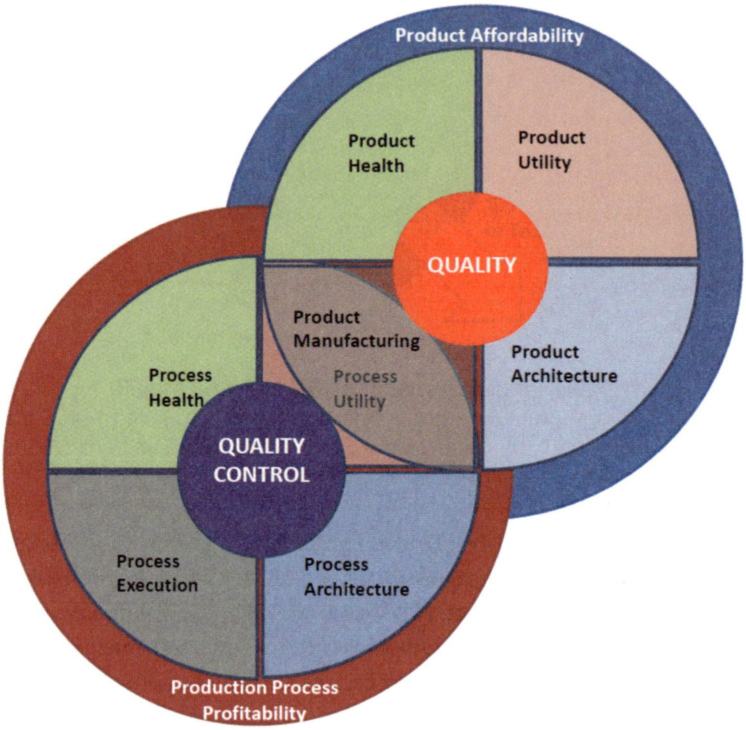

Figure 5.6 The fostering relationship between production process and product.

For the product realization process, considering the natal relationship the process has with the production process and the product, achievements of product quality targets during realization phases depend on realization process manageability and improvability or the joined fit characteristic of quality improvement as shown in Figure 5.7.

To achieve product quality, different approaches for quality improvement and quality management could be followed. The success of a specific approach will depend on the realization process, the production process, the management style, and corporate culture. In the following, two main auto industry quality improvement approaches are discussed according to their chronological development.

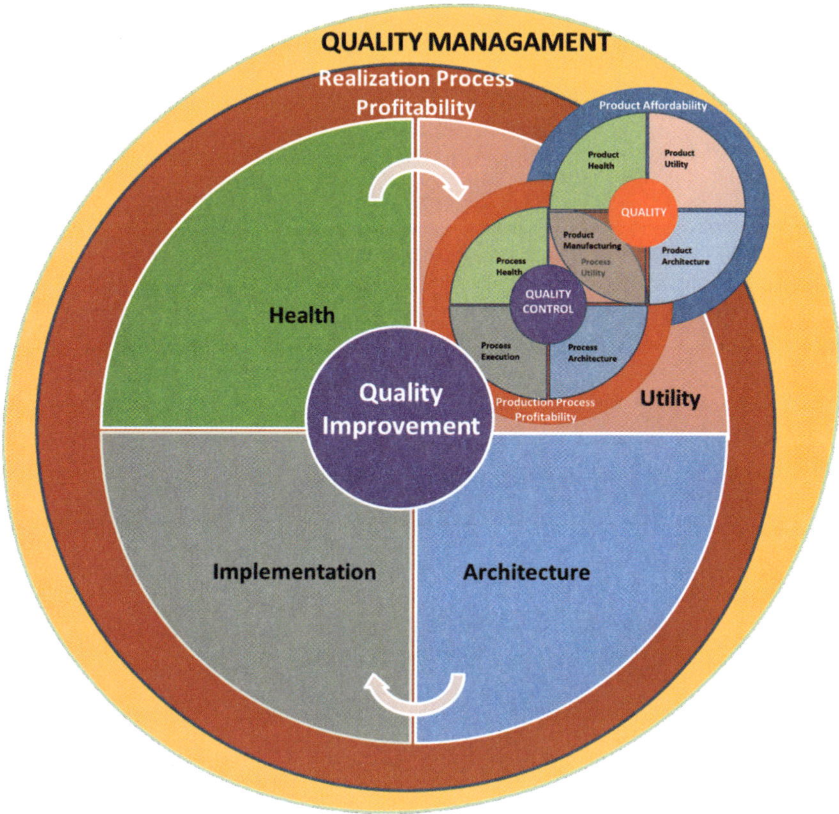

Figure 5.7 Quality improvement through quality management.

5.3.1 Continuous Improvement Approach

In 1945, Taiichi Ohno steered Toyota Production System (TPS) manageability and improvability efforts for improving product quality. As discussed in Chapter 1, he was reassigned from Toyoda Cotton Spinning and Weaving to Toyota Motor for implementing the JIDOKA and just-in-time (JIT) concepts in vehicle mass production [5-12]. Accordingly, TPS became one of the early production processes with quality improvement systems. As a result of studying the TPS, the lean production and lean thinking approaches based on the TPS philosophy, tools, and practices were founded [5-13], [5-14]. Due to the TPS success, lean thinking principles and tools have been adopted and implemented by other industries [5-15], [5-16]. Both TPS and lean production approaches integrate continuous production process improvability with manageability as a production process fit characteristic to achieve product quality, as illustrated in Figure 5.8.

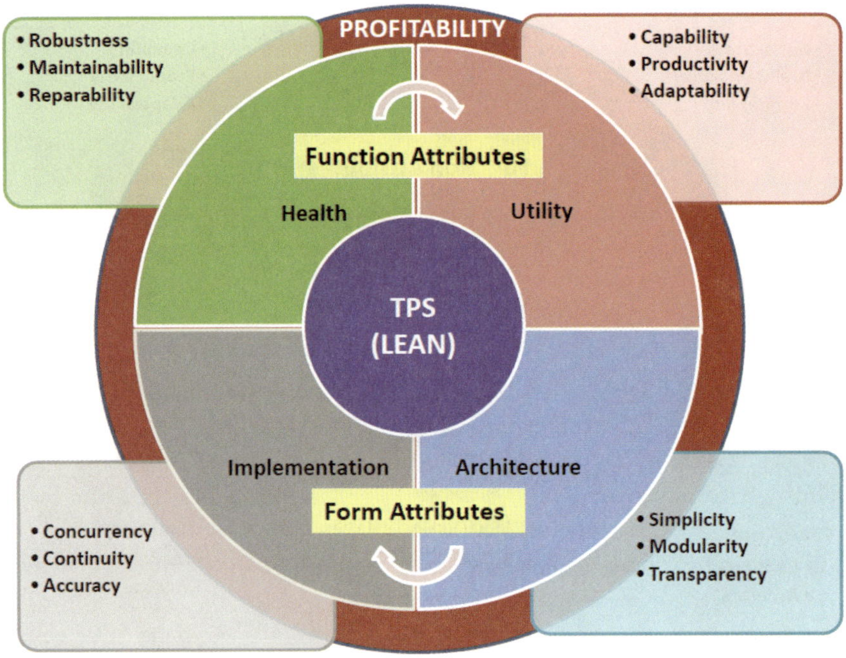

Figure 5.8 TPS and lean thinking joint process manageability and improvability.

For TPS, the "JIDOKA" concept of automation and quality, "JIT" production, removing wastes "MUDA," responding to customer demand as "respect for people," and continuous improvement form the foundations of process manageability and improvability to achieve product quality. Figure 5.9 illustrates the close alignment of these TPS principles with process form, function, and fit characteristics.

Being a derivative of TPS, lean production uses "value" as defined by the customer, "value streaming" in process architecture, the concept of "pull" production, "flow" of nothing but value through eliminating all wastes, and striving for "perfection" for process manageability and improvability to achieve product quality. Figure 5.10 shows the close alignment of the lean principles with process form, function, and fit characteristics.

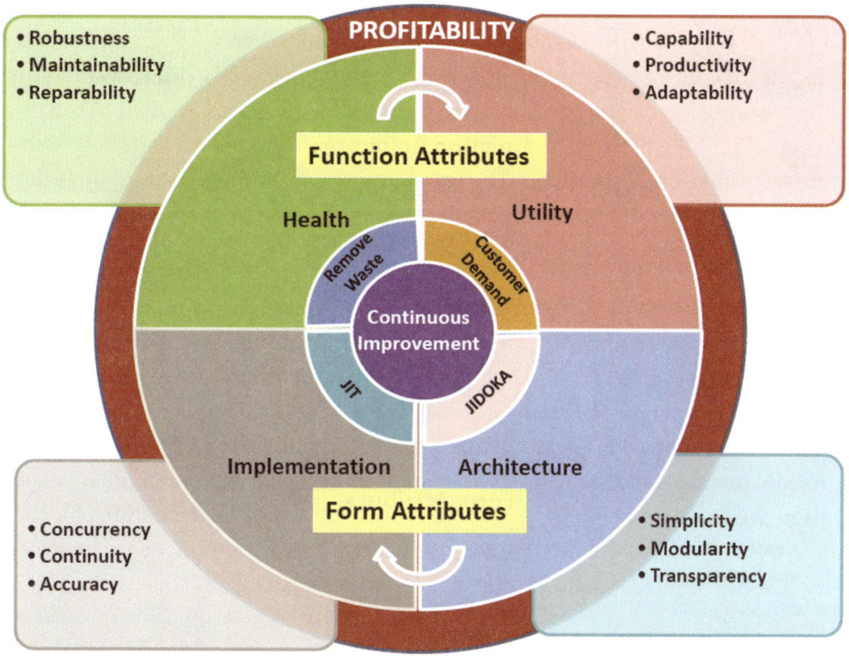

Figure 5.9 Alignment of TPS principles with process form, function, and fit characteristics.

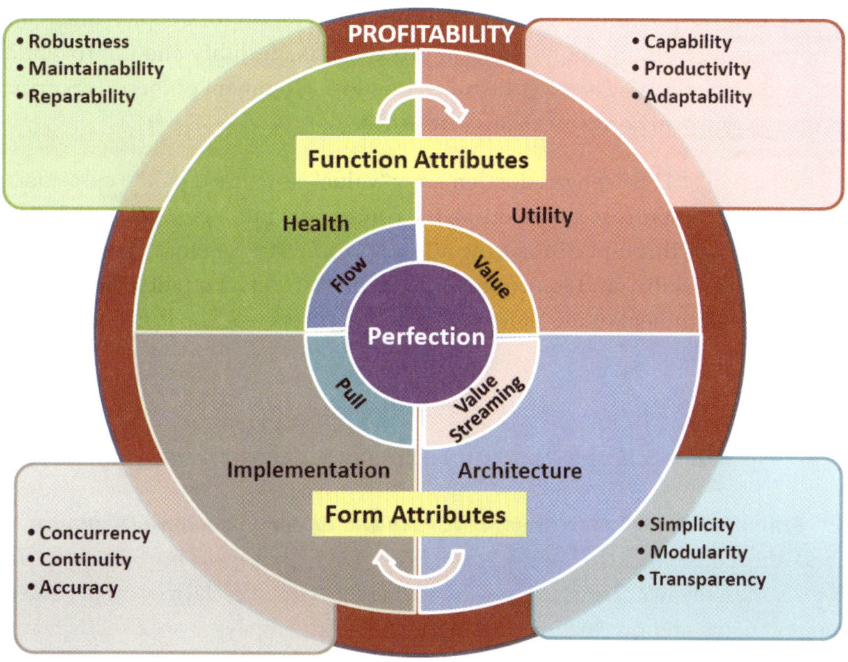

Figure 5.10 Alignment of lean principles with process form, function, and fit characteristics.

In addition to the TPS and lean principles, several tools and techniques have been developed and introduced such as fishbone diagram, Kaizen, seven wastes, standardized work, 5S, visual control, error proofing, and many others [5-17]. While traditionally TPS and lean thinking have been mainly developed and successfully applied to the production process, the same principles and many of the tools can be easily applied to the full product realization process as well. The following three simple examples illustrate the adoption and implementation of three tools for the realization process during product and production process inception, conception, and maturation phases or subphases.

5.3.1.1 Example 5.1: Value Stream Mapping: Value stream mapping is used in manufacturing as a process representation chart for waste minimization. During inception phases or subphases, value stream mapping can be adapted to assure the streaming of product characteristics, attributes, criteria, and targets as deciphered from the VOC (value) to all subsystems and components technical specifications. Figure 5.11 shows the value stream mapping for Safety characteristics and crushability attribute in a vehicle realization process. As shown in Figure 5.11, the deformation specifications for the front-end subsystem and consequently the longitudinal rail as a component are specified based on the full vehicle frontal impact crushability target [5-18].

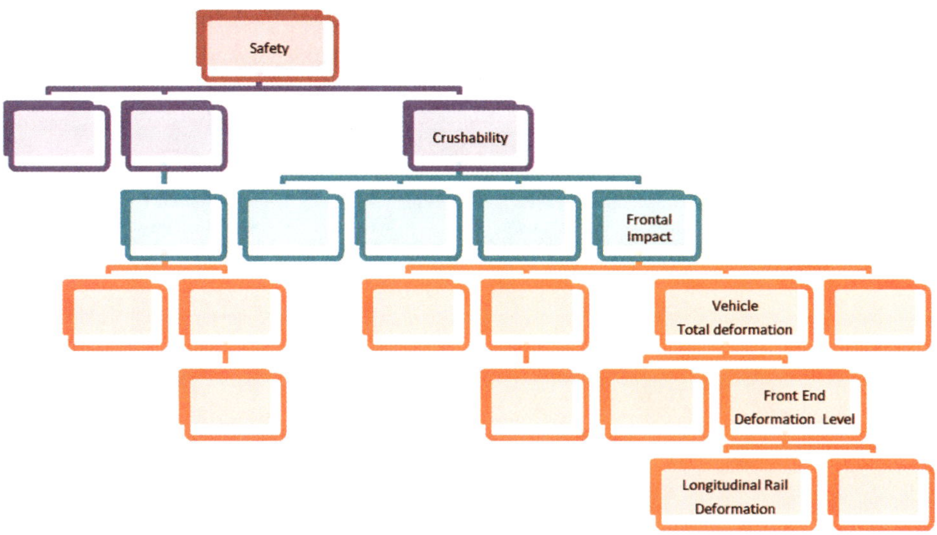

Figure 5.11 Vehicle safety value stream mapping example.

5.3.1.2 Example 5.2: Fishbone Diagram (Cause and Effect or Ishikawa Diagram): The cause and effect diagram also known as the fishbone diagram was introduced in the 1960s by Kaoru Ishikawa. The diagram shows the causes of a certain event or condition. For manufacturing and production processes, root–cause analyses are usually conducted using man, material, machine, and methods known as the 4M. Sometimes environment and measurement are added as shown in Figure 5.12 [5-19].

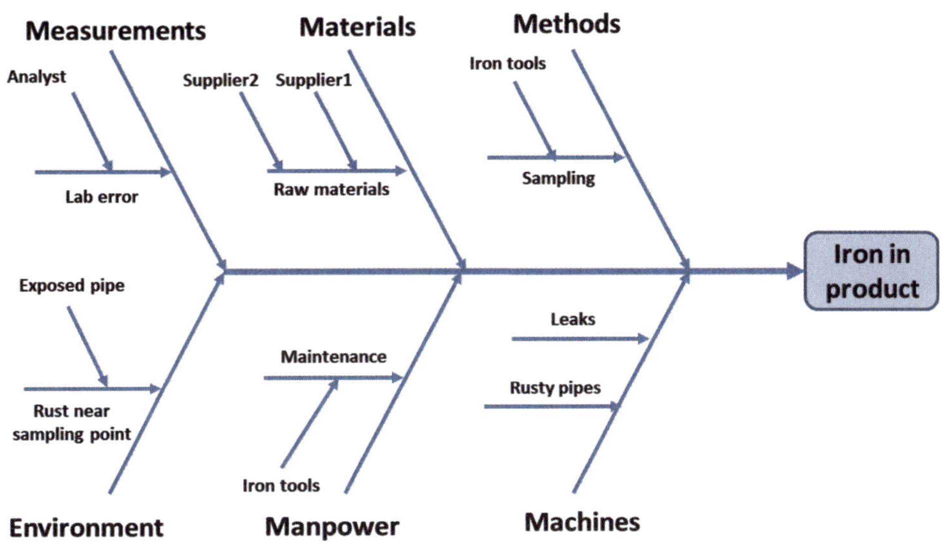

Figure 5.12 Manufacturing fishbone diagram.

To guide and assure the flow of the VOC attributes, the cause and effect diagram could be adapted during conception of different realization phases. Each unachieved target or specification for every customer desired attribute could be tracked down to each subsystem and component. As shown in Figure 5.13, for the front impact safety attribute, the different components of the four subsystem body, interior, power train, and steering are tracked [5-18].

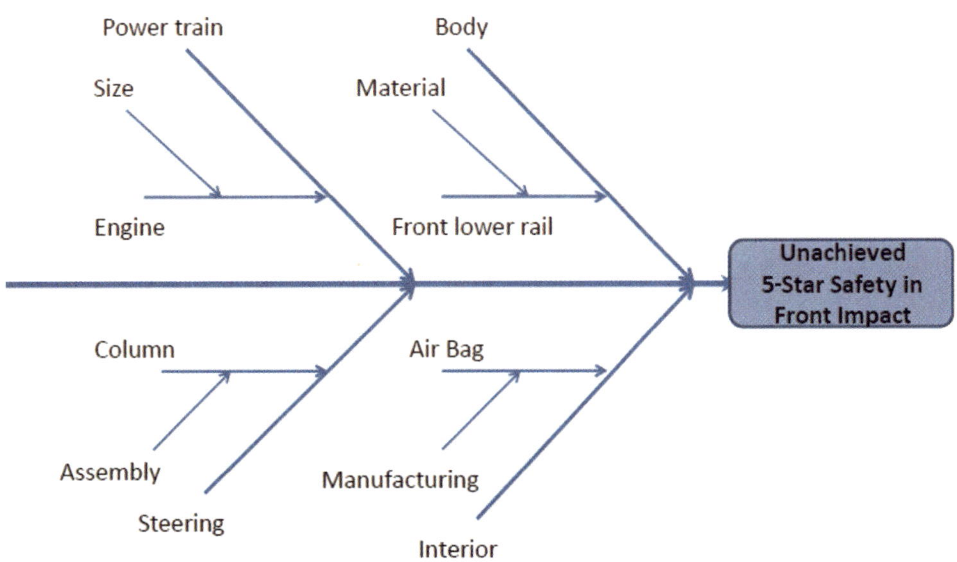

Figure 5.13 Conception fishbone diagram example.

5.3.1.3 Example 5.3: Kaizen: Kaizen is a methodology centered on gathering suggestions, and holding periodic meetings, following the Deming Quality Cycle. During a Kaizen event, a team may take up to five days to identify and resolve an issue. Kaizen methodology could be adapted during a maturation phases or subphases, to assure that unachieved desired characteristic or attribute such as safety or reliability is pulled and addressed, as shown in Figure 5.14 [5-18].

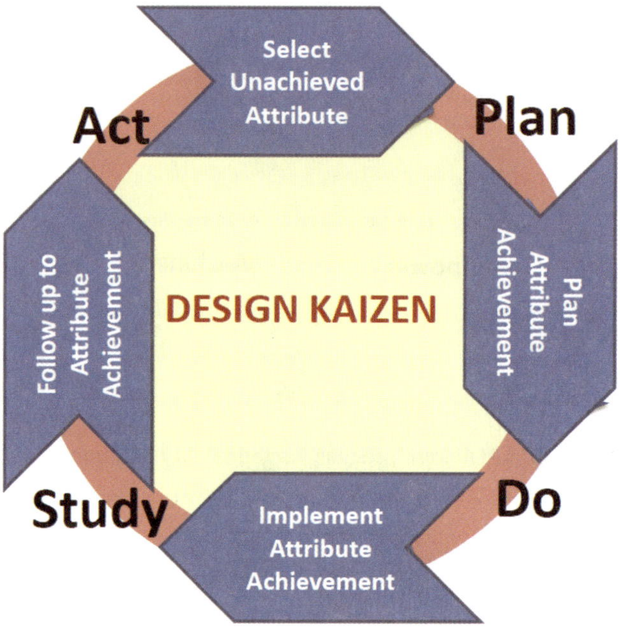

Figure 5.14 Maturation Kaizen event example.

5.3.2 Planned and Structured Approach

In the late 1980s, experts from the three U.S. automobile manufacturers—Ford, GM and Chrysler—started the development of advanced product quality planning (APQP). The APQP was developed to understand customers' demands, plan, assure, manage, improve, and control product or service quality throughout the supply chain [5-20]. To achieve these objectives, APQP integrates production process improvability with manageability during the product realization process as a process fit characteristic to achieve and control product quality, as illustrated in Figure 5.15.

For APQP to achieve product quality targets based on the VOC and assure "quality control" during production, "advanced planning" to meet the customers' expectations, continuous "cross-functional" team monitoring, and resolution of issues through "structured reviews" form the foundation of process manageability and improvability to achieve product quality. Figure 5.16 shows the close alignment of this foundation with process form, function, and fit characteristics.

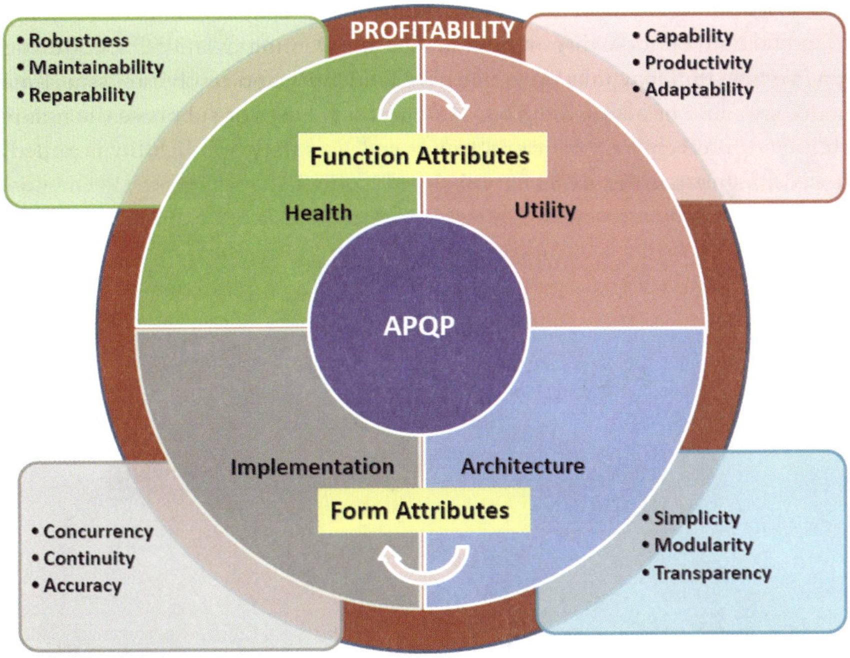

Figure 5.15 APQP joint process manageability and improvability.

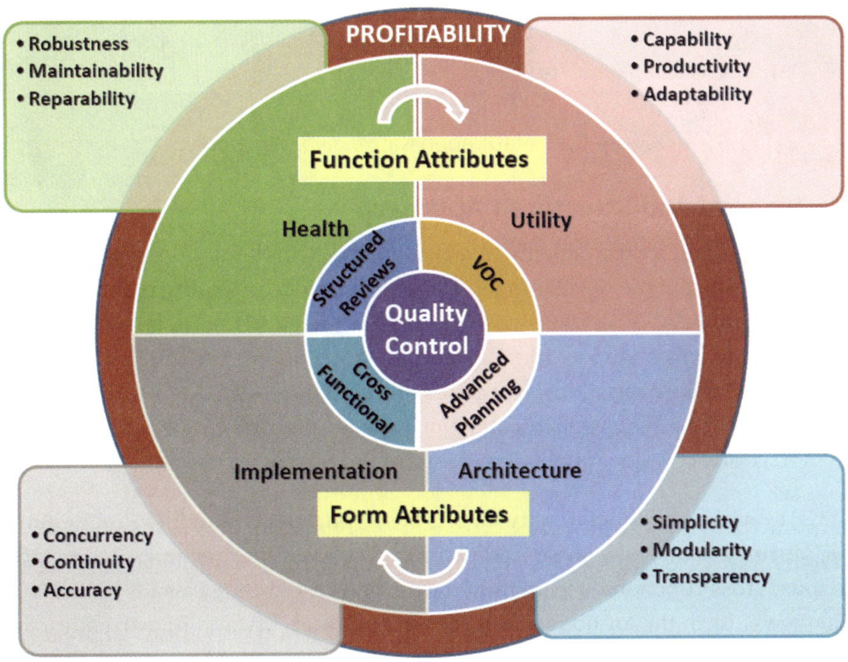

Figure 5.16 Alignment of APQP foundation with process form, function, and fit characteristics.

In 2002, as a result of the success in addressing quality issues, integrating the VOC, assuring and managing quality throughout the supply chain the APQP became an integral element of the automotive quality management system (QMS) International Standard, ISO/TS 16949:2002. The QMS defines the interactions between the activities for managing responsibility, resources, improvement, and the product realization process [5-21], [5-22]. Consequently, for achievements of product quality targets, the relationship among the product realization process, the product, and the production process can be illustrated as shown in Figure 5.17. In the coming chapters, the product realization process phases will be further discussed following the QMS planned and structured approach.

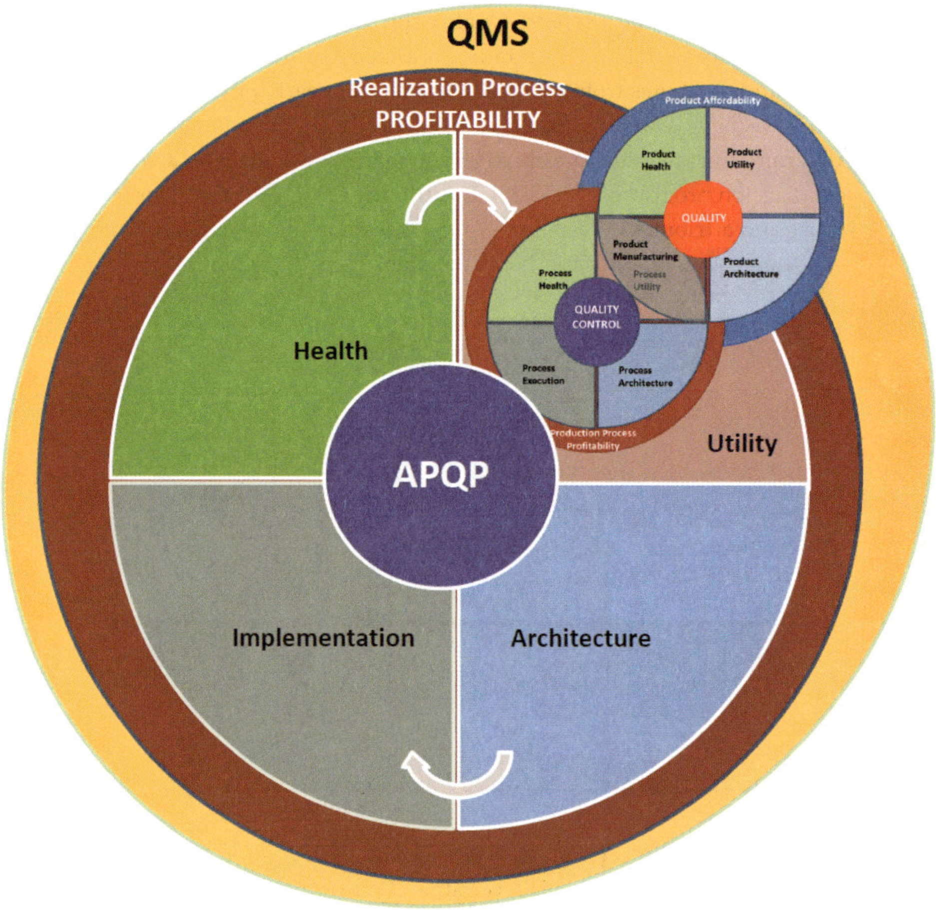

Figure 5.17 QMS and APQP quality improvement through quality management.

Among the key contributions of the APQP is the production part approval process (PPAP). PPAP provides a well-structured and manageable process to regulate approval of products, parts, components, and services for planning, managing, assuring, and achieving quality targets throughout the supply chain. This includes all materials and parts considered during advanced quality planning process. Accordingly, PPAP ensures that all suppliers comply with the design specification and can show documentary evidence that the level of compliance can be maintained during production and product life cycle.

To achieve process improvement and product quality targets, through TPS/lean continuous improvement approach or the APQP planned and structured approach, several methodologies, techniques, quality management, and processes improvement tools are utilized. In the following, selected examples of the methodologies and tools that will be cited in the coming chapters are summarized.

5.3.2.1 Quality Function Deployment: Quality function deployment (QFD) is a structured methodology for managing, facilitating, and deciphering the desired product characteristics and attributes from the VOC starting from early product realization planning phases. QFD's available quality management tools include affinity diagram, relations diagram, tree diagram, matrix diagram, matrix data analysis, arrow diagram, and process decision program chart [5-23]. By utilizing these QFD tools along with the characteristics and attributes driven house of quality template, illustrated by Figure 5.18, customer-driven product specifications can be established.

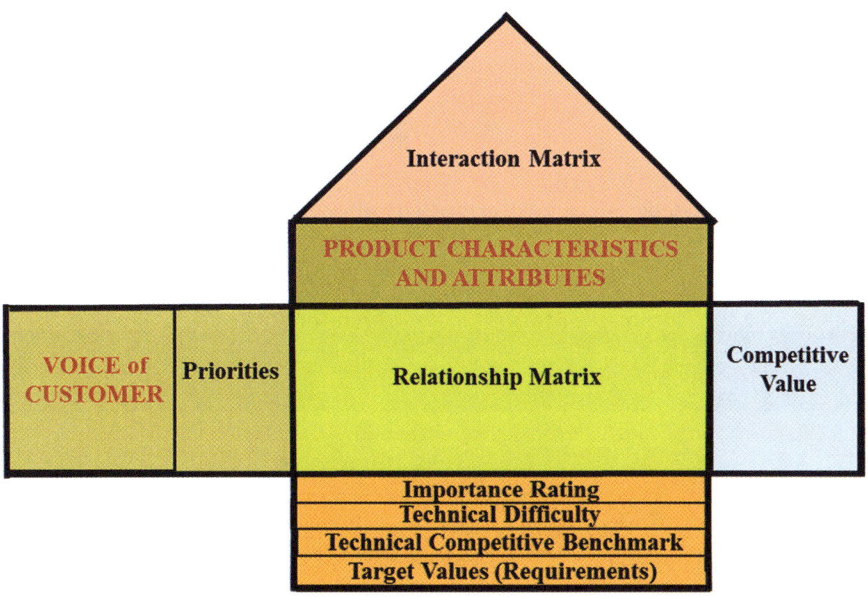

Figure 5.18 Characteristics and attributes-driven house of quality template [5-23]. (Reprinted with permission from ASQ.)

5.3.2.2 Process Methodologies and Tools: In addition to many process methodologies and tools, such as the characteristics matrix and production control plan [5-18], the following are illustrative examples.

5.3.2.2.1 Process Capability: During product and production process virtual and physical realizations, process capability studies must be performed. To ensure that the realized product is within the realized production process capability, the process "capability indices" are developed. For example, two of the indices developed for univariate normally or approximately normally distributed stable processes in statistical control are [5-24], [5-25], [5-26], [5-27], [5-28]

$$C_P = (\text{USL} - \text{LSL})/6\sigma$$

and

$$C_{PK} = \min(C_{PU}, C_{PL})$$

where USL = upper specification limit, LSL = lower specification limit, σ = process standard deviation, and μ = process mean:

$$C_{PU} = (\text{USL} - \mu)/3\sigma, C_{PL} = (\mu - \text{LSL})/3\sigma$$

where C_P is considered a measure for process inherent capability or potential since the index is independent of the process mean. Subsequently, the index C_{PK} was developed to take into account the process mean. For a normally distributed process with centered mean, $C_{PK} = C_P$. The process is considered capable when C_{PK} or C_P is greater than or equal to 1.

To reliably predict or estimate the capability of a process, the data should have sufficient sampling points from a stable process in statistical control with normal or approximately normal distribution [5-24], [5-25], [5-26], [5-27], [5-28]. Table 5.1 provides samples of out of specifications parts per million for process capability index between 0.5 and 2.0. As shown from the table, when designing for process capability, an index value of 1.5 or above should be targeted. To demonstrate the effectiveness of using process capability index, in selecting suppliers and manufacturing equipment, the following example is considered.

Table 5.1 The basic structure of the stage-gate model

Process Capability Index	Out of Specification (Parts per Million)
0.50	133,600.00
1.00	2,700.00
1.50	6.80
2.00	0.0018

5.3.2.2.1.1 Example 5.4: In addition to continuous process and product improvements, process capability analyses can also be used to identify the capabilities of different suppliers. To compare the process capability of two fictitious suppliers, the C_P and C_{PK} for each process histogram are shown in Figure 5.19 [5-29].

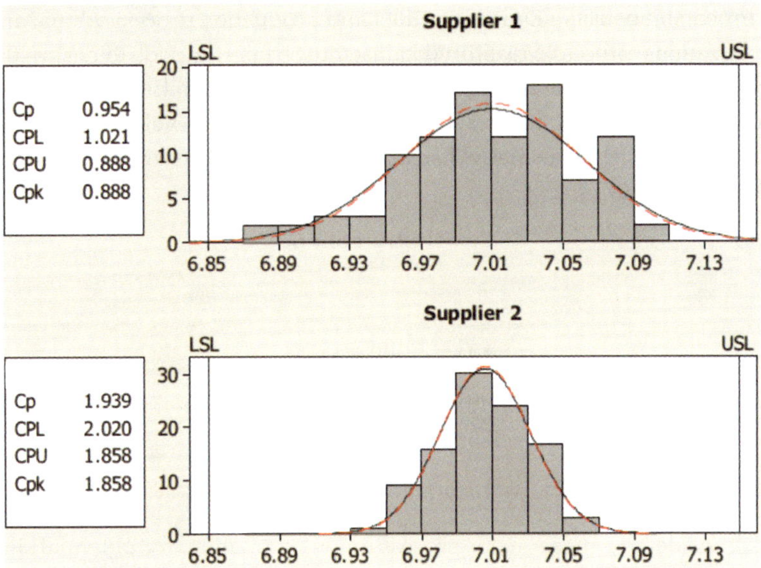

Figure 5.19 C_P and C_{PK} for two histograms of different suppliers' processes [5-29].

As shown in Figure 5.19, for specification of 7+/−0.15, the 6σ process spread of supplier 1 process (having a standard deviation of approximately 0.05) would fit within the specification tolerance approximately one time (C_p = 0.954). For the same specification, the 6σ process spread of supplier 2 process (having a standard deviation of approximately 0.025) would fit within the specification tolerance approximately two times (C_p = 1.939). Therefore, based on C_p only, supplier 2 process has more potential capability than supplier 1 process due to the less inherent variation [5-29]. For considering the location of each distribution within the tolerance window, C_{pk} index could also be calculated [5-29].

5.3.2.2.2 Failure Mode and Effect Analysis: During product and production process virtual and physical realizations, failure mode and effect analyses are conducted by identifying likelihood of a failure mode occurrence, the severity, and ease of detection.

The occurrence rating (OCC), severity rating (SEV), and detection rating (DET) are evaluated according to the following scale:

Occurrence rating (OCC):	1 = not likely, 10 = very likely
Severity rating (SEV):	1 = not severe, 10 = very severe
Detection rating (DET):	1 = easy to detect, 10 = not easy to detect

As displayed by the failure mode and effect analysis (FMEA) example shown in Figure 5.20 [5-30], for each failure mode, a risk priority number (RPN) is calculated as:

$$RPN = SEV \times OCC \times DET$$

Action plans should be created for high RPN to improve the process within the specified time. Different methodologies could be used to reduce the probability of occurrence, severity, and to improve detection through process revisions and error proofing. As illustrated in Figure 5.20, the highest RPN is recalculated after the action plan is implemented to validate improvements. RPN reduction is continued until all potential failure modes are addressed.

POTENTIAL
FAILURE MODE AND EFFECTS ANALYSIS
(PROCESS FMEA)

FMEA Number 1450 (1)
Page 1 of 1

Item Front Door L.H./H8HX-0000-A (2)
Process Responsibility Body Engineering (3)
Prepared by J. Ford - X6521 - Assy Ops (4)

Model Year(s)/Program(s) 199X/Lion 4dr/Wagon (5)
Key Date 9X 03 01 ER 9X 08 26 Job #1 (6)
Detection
FMEA Date (Orig.) 9X 05 17 (Rev.) 9X 11 06 (7)

Core Team A. Tate Body Engrg., J. Smith - OC, R. James - Production, J. Jones - Maintenance (8)

Process Function (9) Requirements	Potential Failure Mode (10)	Potential Effect(s) of Failure (11) (12)→	Sev	Class	←(13) Potential Cause(s)/ Mechanism(s) of Failure (14)	Occur	←(15) Current Process Controls −Prevention −Detection (16) (17)→	Detec	R.P.N.	←(18) Recommended Action(s) (19)	Responsibility & Target Completion Date (20)	Action Results (22)				
												Actions Taken (21)	Sev	Occ	Det	R.P.N.
Manual application of wax inside door	Insufficient wax coverage over specified surface	Deteriorated life of door leading to: ■ Unsatisfactory appearance due to rust through paint over time ■ Impaired function of interior door hardware	7		Manually inserted spray head not inserted far enough	8	Visual check each hour - 1/shift for film thickness (depth meter) and coverage	5	280	Add positive depth stop to sprayer	MFG Engrg 9X 10 15	Stop added, sprayer checked on line	7	2	5	70
To cover inner door, lower surfaces at minimum wax thicness to retard corrosion										Automate spraying	Mfg Engrg 9X 12 15	Rejected due to complexity of different doors on same line				
					Spray head clogged - Viscosity too high - Temperature too low - Pressure too low	5	Test spray pattern at start-up and after idle periods, and preventative maintenance program to clean heads	3	105	Use Design of Experiments (DOE) on viscosity vs. temperature vs pressure	Mfg Engrg 9X 10 01	Temp and press limits were determined and limit controls have been installed - control charts show process is in control Cpk ~ 1.85	7	1	3	21
					Spray head deformed due to impact	2	Preventative maintenance program to maintain heads	2	28	None						
					Spray time insufficient	8	Operator instructions and lot sampling (10 doors/ shift) to check for coverage of critical areas	7	392	Install spray timer	Maintenance 9X 09 15	Automatic spray timer installed - operator starts spray, timer controls shut-off - control charts show process is in control Cpk ~ 2.05	7	1	7	49

SAMPLE

Figure 5.20 FMEA sample [5-30].

5.4 Learning Outcomes from Product Quality and Process Improvability

The desired learning outcomes from product quality and process improvability can be categorized as follows.

5.4.1 Knowledge

Product quality is the peak summative fit characteristic that reflects customers' perceptual level of satisfaction with product form and function characteristics and attributes. Without attention to details in setting and achieving the desired initial and long-term expectations, three producers' gaps in built-in product form and function characteristics and attributes may result. These gaps are the producers' perceptual VOC gaps, producers' development gaps, and production variation gaps. During product evaluation or usage, customers' emotional responses due to expectations, importance, and preferences transform these cognitively perceived gaps into corresponding customers' perceptual quality gaps. Therefore, quality gaps are ultimately the results of the customer's emotional response to the differences between customer expectation or imaginative realization and cognitive evaluation or receptive realization of the product form and function characteristics and attributes.

Modeling the losses associated with the emotional response to the three cognitively perceived gaps can yield some useful insights. For the surprising elements of the producers' G_p and G_d gaps, customers' reactions can be exponential for both positive and negative gaps. The associated losses being proportional to the reaction can be exponentially modeled. As these reactions are usually associated with initial product evaluation, the associated losses are mainly producers' losses in the form of potential sales. For the nonsurprising elements of the producer's G_p and G_d gaps, the reactions can be proportional to both positive and negative gaps. Accordingly, the associated losses can be linearly modeled. As some of these reactions are associated with initial product evaluation, the associated losses could also be producer's losses in the form of potential sales. For unachieved G_d and production variation G_v gaps, losses are usually modeled as quadratic functions for both positive and negative gaps. The associated losses usually appear during long-term customer's usage. Therefore, these losses are mainly customer's losses. However, customer's losses ultimately cause producer's losses in the form of customers' loyalty and market share.

Considering the natal relation the production process has with the product, achievements of product quality targets during production depend on production process manageability and improvability, or simply quality improvement and control. To achieve product quality, different approaches for quality management and improvement could be followed. The two main approaches followed in the auto industry are the TPS/lean continuous improvement approach and the APQP/QMS planned and structured approach. The TPS/lean approach focuses on continuously improving the production process, but can be easily extended to the full product realization process. The APQP/QMS planned and structured approach focuses on planning and structuring the product realization process, but can be easily extended to continuously improve the production process.

5.4.2 Skills

The skills needed for setting and achieving quality targets depend on the product realization phase. Understanding the customer and the ability to decipher expected and desired product characteristics and attributes are the skills needed at the early product inception phase. Recognizing the product gaps for being positive or negative, and the type of customer behavior for being linear or exponential are the skills needed for guiding product planning phase. Abilities to close product gaps are the skills needed during conception phases and virtual realization. Finally, attention to detail and anticipation of quality issues are the skills needed during maturation phases and physical realization.

5.4.3 Paradigm/Attitude

While poor quality could cause losses to the customers and society, the highest share of these losses will ultimately become the producer's losses in the form of potential sales, customers' loyalty, and market share.

5.5 References

5-1. Castro, J. 1989. "The Quest for Quality in U.S. Goods: Making It Better," Times: Tokyo.

5-2. Ullman, D. 2009. "The Mechanical Design Process." 4th Ed. McGraw-Hill Publishing Company: New York.

5-3. http://www.iso.org/iso/catalogue_detail.htm?csnumber=15570.

5-4. Kano, N., Seraku, N., Takahashi, F., and Tsuji, S. 1984, "Attractive Quality and Must-Be Quality, Hinshitsu." *Journal of the Japanese Society for Quality Control,* 39–48.

5-5. Berger, C., Blauth, R., Boger, D., Bolster, C., Burchill, G., DuMouchel, W. et al. "Kano's Methods for Understanding Customer-defined Quality." *Center for Quality Management Journal,* 4 Fall 1993:3–36, 2011.

5-6. Matzler, Kurt, and Hans H. Hinterhuber. "How to Make Product Development Projects More Successful by Integrating Kano's Model of Customer Satisfaction into Quality Function Deployment." *Technovation* 18(1):25–38, 1998.

5-7. http://www.jdpower.com/sites/default/files/2014087_IQS.pdf.

5-8. Taguchi, G. 1986. Introduction to Quality Engineering: Designing Quality into Products and Processes, Asian Productivity Organization, UNIPUB/Kraus International: White Plains, NY.

5-9. Phadke, S. M. 1989. Quality Engineering Using Robust Design, Prentice Hall: Englewood Cliffs, NJ.

5-10. Wysk, R., Niebel, B., Cohen, P., and Simpson, T. 2000. Manufacturing Processes: Integrated Product and Process Design. McGraw Hill: New York.

5-11. http://autos.jdpower.com/ratings/2014-Vehicle-Dependability-Study-Press-Release.htm.

5-12. Mass, W., and Robertson, A., "From Textiles to Automobiles: Mechanical and Organizational Innovation in the Toyoda Enterprises," Business and Economic History, 25(2):1895–1933, 1996.

5-13. Womack, J. P., Jones, D. T., and Roos, D. 1990. "The Machine That Changed the World: The Story of Lean Production." Rawson and Associates: New York.

5-14. Womack, J. P., and Jones, D. T. 1996. Lean Thinking: Banish Waste and Create Wealth in You Corporation. Simon & Schuster: New York.

5-15. Curt Hibbs, C., Jewett, S., and Sullivan, M. 2009. "The Art of Lean Software Development: A Practical and Incremental Approach," O'Reilly Media, California.

5-16. Chalice, R. 2007. Improving Healthcare Using Toyota Lean Production Methods: 46 Steps for Improvement, 2nd Edition. American Society for Quality, Quality Press: Milwaukee.

5-17. http://www.leanproduction.com/top-25-lean-tools.html.

5-18. M. El-Sayed. *Implementation of Lean Tools and Methodologies in Design*, ASME Paper No. IMECE2012-89030, Houston, Texas, November 2012.

5-19. http://asq.org/learn-about-quality/cause-analysis-tools/overview/fishbone.html.

5-20. Chrysler Corporation, Ford Motors Company, and General Motors Corporation. *Advanced Product Quality Planning (APQP) and Control Plan Reference Manual*, Issued June 1994.

5-21. ISO. 2003. ISO/TS 16949:2002 Implementation Guide, CQI-7, Automotive Industry Action Group: Michigan.

5-22. ISO. 2009. Technical Specification ISO/TS 16949, Quality management systems—Particular requirements for the application of ISO 9001: 2008 for Automotive Production and Relevant Service Part Organizations, Third Edition. ISO, Available from: http://www.iso.org/iso/catalogue_detail?csnumber=52844.

5-23. Revelle, Jack B. (Quality Press, 2004) Quality Essentials: A Reference Guide from A to Z. Benchmarking: House of Quality Image.

5-24. Kotz, S., and Johnson, N. 1993. Process Capability Indices. Chapman & Hall: New York.

5-25. Chang, P. and Lu, K., PCI calculations with any shape distribution with percentile, *QWTS*, 110–114, 1994.

5-26. Kotz, S. and Lovelace, C. 1998. Process Capability indices in Theory and Practice. Arnold: London.

5-27. Kotz, S., and Johnson, N. Process Capability Indices—A Review, 1992–2000. *Journal of Quality Technology* 34(1):2–53. 2002.

5-28. Ramanathan, T., Dharmadhikari, A., and Abraham, B., Nonparametric Capability Indices, *Economic Quality Control* 18(1):31–41, 2003.

5-29. Bauer, D., "Statistical Capability Analysis for Geometric Tolerances Using the MMC Modifier," SAE Technical Paper 2009-01-1546, 2009, doi:10.4271/2009-01-1546.

5-30. Society of Automotive Engineers (SAE), "Potential Failure Mode and Effects Analysis in Design (Design FMEA), Potential Failure Mode and Effects Analysis in Manufacturing and Assembly Processes (Process FMEA), and Potential Failure Mode and Effects Analysis for Machinery (Machinery FMEA)," J1739_200006, June 2000.

Chapter 6
Front End of Innovation in Vehicle Realization

"Every improvement or innovation begins with an idea. But an idea is only a possibility—a small beginning that must be nurtured, developed, engineered, tinkered with, championed, tested, implemented and checked—ideas have no value until they are implemented."

—Alan Robinson and Dean Schroeder

6.1 Introduction

In a competitive global market, one of the keys to success is sustained product or service innovations. In general, sustained product innovations are faced with two issues. The first issue is the level of desired innovation. The second issue is the structure of the early stages or the front end of innovation. In the auto industry, with a highly competitive global market, the same product innovation issues are facing many organizations.

The levels of innovation at any organization depend on the internal and external drives and constraints. To achieve certain levels of innovation, specific goals must be established. Therefore, understanding innovation levels is necessary for setting these goals and assessing associated risks.

Due to its effect on product innovation, the front end of the product realization process is very critical for organizational competitiveness and success. Because of the fuzzy nature of product ideas, customer requirements, and market needs at the front end, the term "fuzzy front end of innovation," or simply "fuzzy front end," is commonly used in the literature to describe the front end of new product development activities [6-1], [6-2],

[6-3], [6-4], [6-5], [6-6], [6-7]. Despite its importance in keeping the competitive edge of the organization, the fuzzy front end is usually the least understood, hardest to manage, and most poorly executed phase of the product development process.

In this chapter, to address the two issues facing sustained product innovation, the innovation levels and the front end of innovation in the context of product realization are studied. The main objectives are redefining these levels and understanding sustained innovation. In addition, some relevant models and a new approach for innovation during the preinception phase of vehicle realization will be discussed.

6.2 Innovation Levels

Despite many research efforts [6-1], [6-2], [6-3], [6-4], [6-5], [6-6], [6-7], [6-8], [6-9], [6-10], [6-11], [6-12], [6-13], the term innovation has no universal definition. The reason is that innovation describes a context-dependent, multidirectional, and multilevel endeavor. For example, in the context of technology, innovation could mean improved technique or new application. In the context of product realization, innovation could mean new form or improved function. In the context of marketing, innovation could mean market forecast and/or market strategy. In the context of organizational business, it could include marketing and/or technology innovations. In addition, the levels of innovations could be low or high for both technology and marketing.

With two levels of low and high for each of the technology and marketing innovation dimensions, there are four quadrants as shown in Figure 6.1 [6-14]. The lower left quadrant characterizes low or incremental marketing and technology innovations. The lower right quadrant represents incremental marketing innovation and high technology innovation. The upper left quadrant represents the incremental technology innovation and high market innovation. Finally, the right upper quadrant characterizes the high technology and high marketing innovation labeled as "radical" innovation [6-14]. Along with low and high levels of innovation in marketing and technology, the associated uncertainty and risks are also indicated (with the red arrows).

Considering that radical innovation may also include far-reaching, breakthrough, and disruptive innovations that are usually associated with a leap from the current state, it is better to distinguish radical innovations from major innovation. Even with high levels of innovations, a continuation or an extension of a previous innovation state does not necessarily mean breakthrough or disruption. Accordingly, the term radical innovation in Figure 6.1 should be replaced with the term major innovation. As shown in Figure 6.2, radical or breakthrough disruptive innovation is a result of leaping in one or both innovation dimensions.

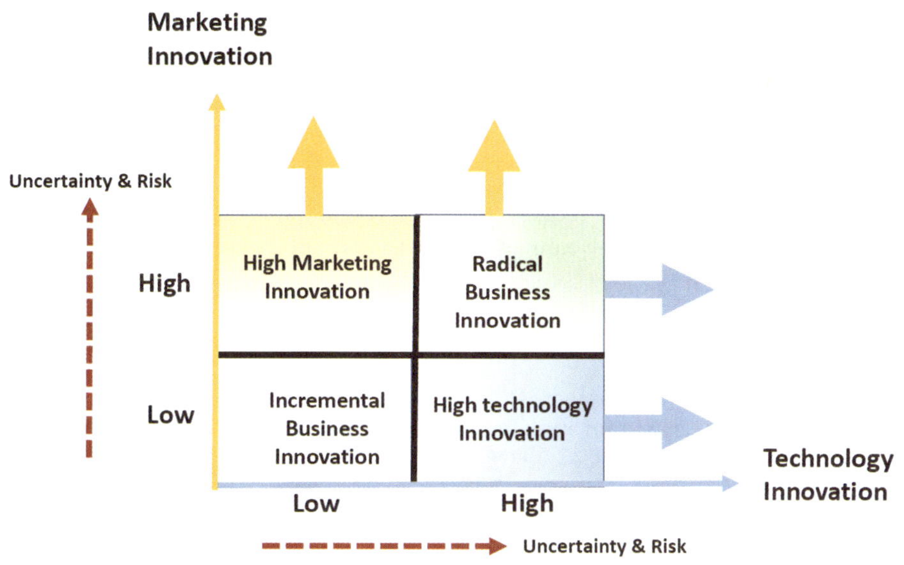

Figure 6.1 Business innovation levels [6-14].

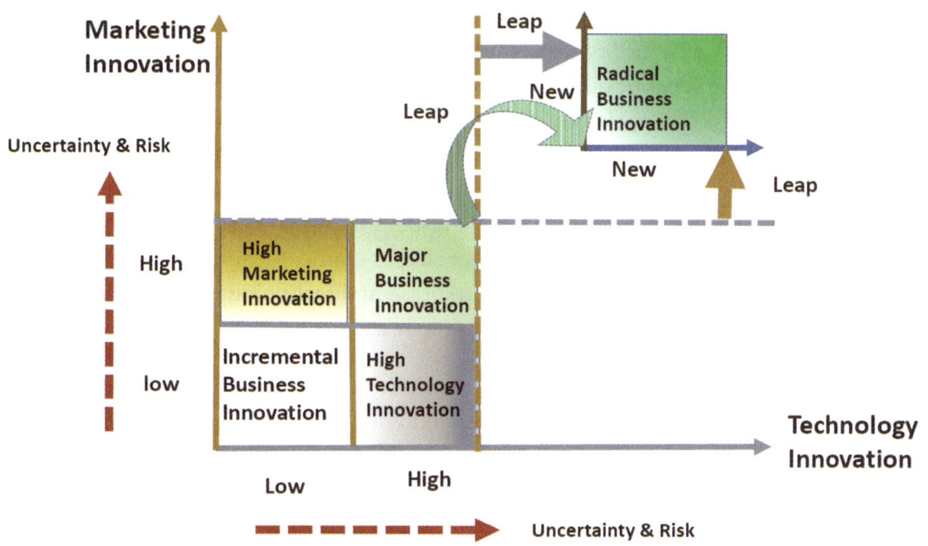

Figure 6.2 Leaps in business innovations.

For the different levels of innovations in the context of marketing, the two dimensions of market demand (creating demand) and market strategy (satisfying demand) are shown in Figure 6.3. The lower left quadrant represents low or incremental marketing innovations. The lower right quadrant represents high market strategy innovation. The left upper quadrant represents high market demand innovation. The right upper quadrant represents major marketing innovation with the high level of forecast and high level of demand innovations. Finally, the radical marketing innovation occurs with leaps in one or both demand and strategy innovations.

Figure 6.3 Marketing innovation levels.

For the technology context, the two dimensions of application (implementation) and the technique (method), can be selected as shown in Figure 6.4. The lower left quadrant represents low or incremental technology innovation. The lower right quadrant represents high application innovation. The left upper quadrant represents high technique innovation. The right upper quadrant represents major technology innovation with the high levels of application and technique innovations. Finally, the radical technology innovation occurs with leaps in one or both application and technique innovations.

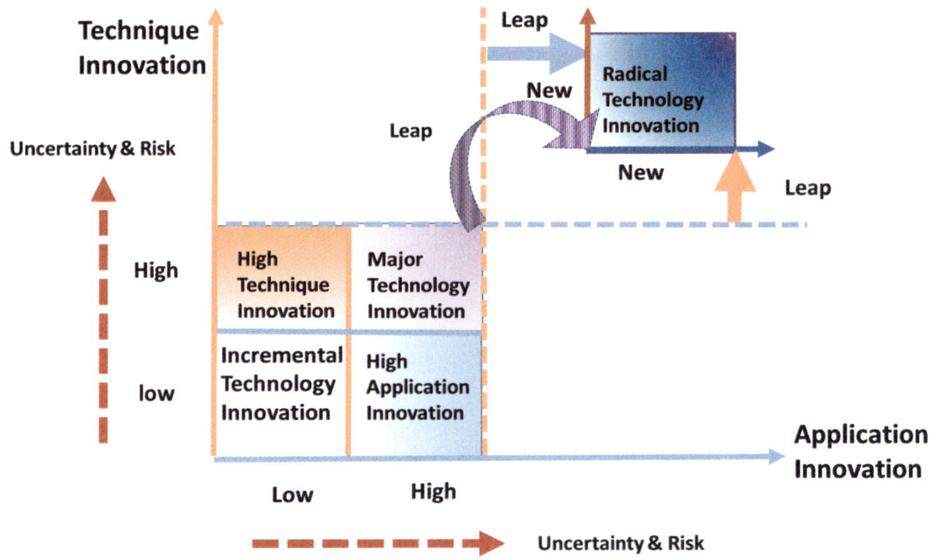

Figure 6.4 Technology innovation levels.

In the product realization context, with the two dimensions of form and function, the different levels of innovations are shown in Figure 6.5. The lower left quadrant represents low or incremental product innovation. The lower right quadrant represents high form innovation. The left upper quadrant represents high function innovation. The right upper quadrant represents major product innovation with the high level of form and high level of function innovations. Finally, the radical product innovation occurs with leaps in one or both form and function innovations.

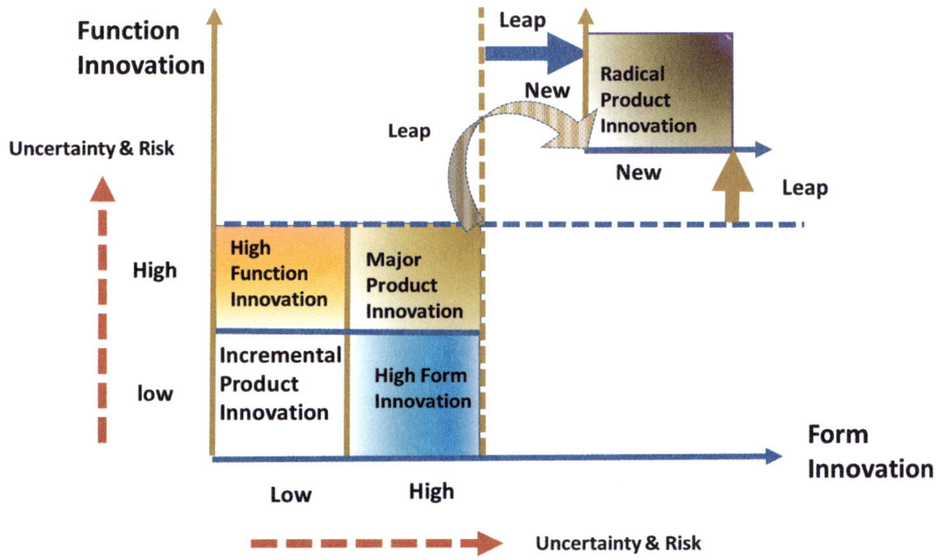

Figure 6.5 Product innovation levels.

In the previous figures, the red dashed arrows illustrate that the uncertainty and risks associated with innovations are usually proportional to the innovation level. Consequently, the higher the level of innovation the higher the uncertainties and risks. Therefore, the highest uncertainties and risks may be associated with radical innovations, and the lowest uncertainties and risks are associated with incremental innovations. However, for a newly introduced unique radical product with no similar form and/or function competitive products, the radical innovation axes would be at the origin and the levels of associated risks would be lower.

The scope and magnitude of product innovations and associated risks depend on many factors such as the type of product, industry, and market. Examples of these factors include the following:

- number of disciplines involved in product development
- effort and time needed for product development
- number of components and subsystems in the product
- number of technologies and degrees of complexity in each subsystem
- type and state of each utilized technology
- required infrastructure for product support and utility
- expected life cycle of the product
- probability and consequences of failure
- supply chain
- external rules and regulations
- established norms and expectations
- market needs and readiness
- customer's usage environment
- customer's financial and emotional investment in the product

6.3 Innovation Levels in the Auto Industry

In industries with products made of multi-subsystems and components such as the auto industry, it is important to make the distinction between the product innovation levels and the innovation levels of the technologies employed in the product, subsystems, and components. It is also necessary to make the distinction between innovations and improvements in products and technologies. For example, not all vehicle and technology improvements are results of innovations and not all innovations cause improvements.

Similar to any product, vehicle innovations have the form and function innovation dimensions, as shown in Figure 6.5. In general, vehicle form innovations attract and please customers, while function innovations excite and thrill them. Both form and function innovations, however, are desired and usually demanded by customers. Unlike marketing innovations, where the risks associated are usually assumed by the organization, some of the vehicle and technology innovation risks may be passed to the customers. Passing innovation risks to the customers, however, is the main cause for loosing customers' loyalty. Because of the well-established customers' expected characteristics, the key to successful vehicle and technology innovations is to minimize the associated risks to the customers. Historically, incremental vehicle and technology innovations with minimum associated risks to customers have shown to be more effective than radical innovations with higher associated customers' risks. This can be illustrated by considering the following case study.

6.3.1 Case Study 6.1: Toyota Camry

A typical case of continued and successful market introductions of vehicle and technology innovations can be deciphered by studying the history of the Toyota Camry. The first generation Toyota Camry, as shown in Figure 6.6, was introduced in Japan in 1980, and to the U.S. market in 1983 as a replacement for the Corona [6-15].

Figure 6.6 First generation Toyota Camry.
(Reprinted with permission from TOYOTA MOTOR SALES, U.S.A., INC. www.toyota-global.com/company/history_of_toyota/75years/vehicle_lineage/car/id60010264/index.html.)

The incremental vehicle innovations and marketing introductions of the Camry are outlined in Table 6.1 [6-15].

Table 6.1 Toyota Camry marketing introduction and vehicle innovation history. (Reprinted with permission from TOYOTA MOTOR SALES, U.S.A., INC. http://pressroom.toyota.com/article_display.cfm?article_id=2493.)

Calendar Year	Introduction/Enhancement
1983	Introduced as replacement for Corona.
1987	Second generation introduced. All Camry vehicles powered by 2.0L DOHC engines. Introduction of All-Track to Camry line. Introduction of Camry wagon. Toyota Motor Manufacturing, Inc. (TMM) announced.
1988	First year of U.S. V6 added to Camry line.
1992	Third generation introduced. New Camry moves to mid-size classification.
1993	Coupe model introduced.
1994	All-aluminum V6 engine introduced.
1995	Minor front and rear fascia changes.
1997	Coupe and Wagon discontinued. Fourth generation introduced.
1998	Camry Solara sport coupe is introduced for '99 Model Year.
2001	Special "Gallery Series" edition available on the Camry LE.
2002	Fifth-generation Camry introduced.
2003	Beginning in January 2003, Camry four-cylinder models sold in California receive engines and automatic transmissions for a Partial Zero Emission Vehicle (PZEV).
2004	Changes to engine and drivetrain for XLE and SE models, Limited Edition Camry.
2005	Five-speed automatic transmission and Vehicle Stability Control (VSC) available for four-cylinder models, ABS standard for all, with minor style freshening.
2007	Sixth-generation Camry introduced. Camry Hybrid is the first Toyota hybrid to be built in the U.S.
2010	2.5-L four-cylinder and six-speed automatic transmission, with restyle front and rear.
2012	Seventh-generation Camry launched.

The incremental vehicle and technology innovations of the Camry resulted in numerous achievements and recognitions as demonstrated in Table 6.2 [6-15].

Table 6.2 Toyota Camry awards and recognitions history. (Reprinted with permission from TOYOTA MOTOR SALES, U.S.A., INC. http://pressroom.toyota.com/article_display.cfm?article_id=2493.)

Calendar Year	Awards/Recognition
1983–1994	Among "Best Buys" by *Consumers Digest*.
1985	"Most Trouble-Free Car in America," *Consumer's Digest*.
1990	*Family Circle* "Family Car of the Year." Named "Best in Class" in Initial Quality by J. D. Power and Associates. Among "Top Ten Models in Initial Quality," J. D. Power and Associates.
1991	Among "Top Ten Models in Initial Quality," J. D. Power and Associates.
1992	Wins *Family Circle* "Family Car of the Year." Among "Top Ten Models in Initial Quality," J. D. Power and Associates. Among "Ten Best Cars," *Car and Driver*.
1992–1993	Camry ranks in Top-10 in Initial Quality, J. D. Power and Associates.
1993	Named "Best in Price Class," J. D. Power and Associates. Among "Ten Best Cars," *Car and Driver*. Named "All Star" by *Automobile Magazine*.
1993–1994	"Family Car of the Year" Sedan Finalist, *Family Circle*.
1994	"Top Ten Models in Initial Quality," J. D. Power and Associates.
1996	"Best Buy in Family Car Segment," *Consumers Digest*. "Best of What's New," *Popular Science*.
1997	Camry is the best-selling car in America for both '97 model and calendar year.
2006	Among "Best Buys," *Consumers Digest* 2006—Among *Consumer Guide's* 2006 Best Buy. *Car Book 2006* "Best Bet."
2007	Camry Hybrid named "Eco-Friendly Car of the Year" by Lifestyle Awards. Camry Hybrid named "Cars.com Lifestyle Award Winner" in the Car of the Year category. "Car of the Year," *Motor Trend.* "Best of the Best," (hybrid version) *The Detroit News.*
2008	Camry named "Best-Selling Automobile" by *Los Angeles Business Journal.* Named top hybrid car for 2008 by *the U.S. News and World Report.*
2009	*The U.S. News and World Report*'s 2009 Best Car for the Money Award. "Best Bets," Cars.com.
2012	Named "Best Resale Value-Mid-Size Car" by *Kelly Blue Book.*

An example of minimizing the vehicle and technology innovations customers' risks, of these marketing introductions, is demonstrated by the introduction of the hybrid powertrain to the U.S. Camry. As shown in Table 6.1, in 2007, the hybrid powertrain, one of Toyota's major technology and vehicle innovations, was introduced to the Camry as the first Toyota hybrid to be built in the U.S. Before its introduction to the U.S. Camry, the hybrid powertrain was continuously improved since its introduction to the Toyota Prius in Japan 1997 [6-15].

Following the principles of continuous improvement, augmented with incremental reliable product and technology innovations, Toyota Motor Company, starting in 1937, became the number one automaker in the world in 2008. Starting with the production of the G1 Truck in 1935 and the first Toyota Motor launched vehicle in 1947, Toyota continued to grow as shown in Figure 6.7 [6-15]. After the production disruptions caused by the 2011 earthquake and tsunami, Toyota's production exceeded the unprecedented 10 million vehicles in 2013 [6-15].

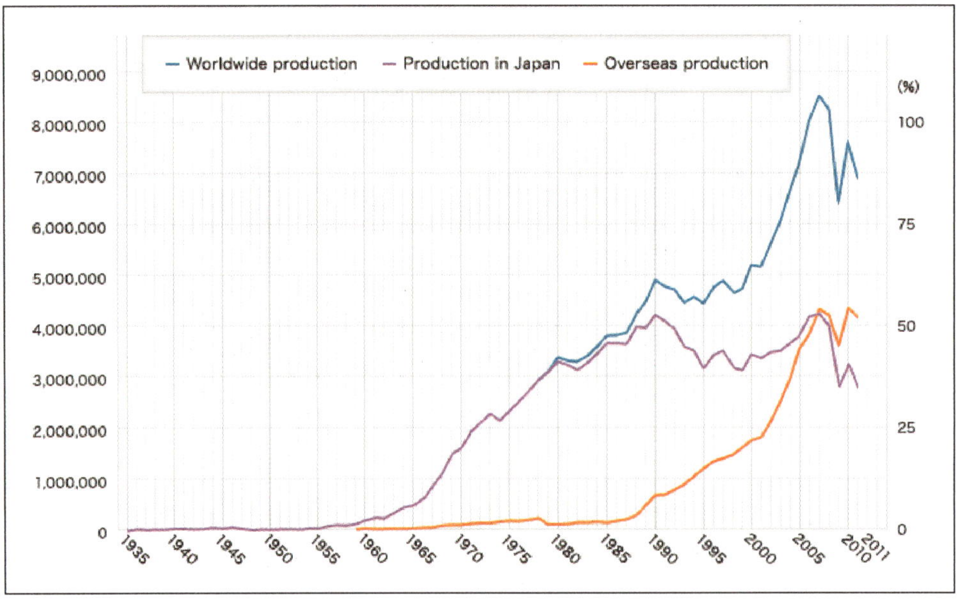

Figure 6.7 Toyota worldwide production history (number of vehicles versus calendar year). (Reprinted with permission from TOYOTA MOTOR SALES, U.S.A., INC. http://www.toyota-global.com/company/history_of_toyota/75years/data /automotive_business/production/production/overview/index.html.)

In general, the levels of technology innovations do not usually translate to the same levels of vehicle innovations. For instance, a radical innovation in airbag technology may cause vehicle safety improvement but does not translate to a radical level of vehicle innovation. In some special cases, depending on the subsystem and its impact on vehicle key characteristics, the level of technology innovation may translate or lead to an equivalent level of vehicle innovation. The development of electric powertrain technology can be considered a major technology innovation, and implementation of this technology in vehicles can also be considered as a major vehicle innovation. Similarly, radical technology innovation could lead, in some cases, to radical vehicle innovation as illustrated by the following GM Autonomy case study.

6.3.2 Case Study 6.2: GM Autonomy

To change the dynamics of auto production and allow fuel cells to compete with the deeply rooted internal combustion engines, in 2002 GM introduced the Autonomy concept vehicle, shown in Figure 6.8 [6-16]. The concept vehicle 6-in-thick rolling chassis, shown in Figure 6.9 [6-17], integrated the fuel cells with the vehicle's powertrain. The radically innovative skateboard-like chassis platform could be attached to a variety of body types.

Figure 6.8 The 2002 GM autonomy concept vehicle [6-16].

Figure 6.9 The 2002 GM autonomy rolling chassis [6-17].

With the radical changes in design, it was predicted by GM executives that "the concept vehicle would cause a quantum leap in vehicle technology and economics not seen since Henry Ford's Model T" [6-18]. However, despite the investment of over a billion dollars, the Autonomy concept vehicle and its follow-up derivatives, such as the 2005 Chevy Sequel, have not made it to mass production due to several hurdles, including the following:

- hydrogen fueling infrastructure
- establishing cost effective mass production processes for fuel cell powertrains
- developing better on-board hydrogen storage technologies

To overcome some of these difficulties in 2013, GM and Honda announced that they would work with other parties to tackle the hydrogen refueling infrastructure issue. The collaborative effort aimed at the mass production of the "next-generation" fuel cell vehicles with new hydrogen storage systems by 2020 [6-19].

As demonstrated by the two case studies, market introductions of sustained long-term incremental vehicle and technology innovations have been traditionally more successful than the introduction of radical vehicle and technology innovations. As shown by the Autonomy case study, the external constraints, in addition to the higher uncertainty and risks associated with radical vehicle and technology innovations, tend to sway the industry toward incremental innovations with a few major innovations when necessary. Therefore, with limited exceptions, incremental innovations tend to be the dominant operational norm in the auto industry. This norm is caused by outside constraints in addition to organizational as well as customers' factors. Some of these factors from an organizational perspective include the following:

- conformity with external rules and regulations
- compatibility with existing infrastructure
- dealing with traditional expectations and established markets
- losing cost benefits of using already existing production infrastructure
- potential losses associated with unknown risks

Some of the factors from the customer's perspective include the following:

- Affordability, usually a key deciding factor for ownership.
- Safety risks are not tolerated.
- Quality, reliability, durability, and dependability are the necessary key characteristics for daily operation.

- Ease of use and interface are desired and relearning is unpopular.
- Usability on existing roads and fueling infrastructure are essentials.

6.4 Front End of Innovation Models

Most organizational front-end activities leading to new product realization are perceptual. Perceptual realizations are in general less crisp than virtual realizations which, in turn, are less concrete than physical realizations. Formulating the desired product's definition using the voice of the customer (VOC), as a key input, requires the mapping from the customer's perceptual domain to the developer's perceptual domain. Considering the fuzzy nature of customer perceptual definitions, the formulated developer's perceptual definition can hardly be crisp. Similarly, for market and technology trends beside uncertainty, all forecasts are perceptual in nature with intrinsic built in fuzziness. In addition, the segmentation factor due to having different pieces in the perceptual domain of different team members causes the collective perception to be extra fuzzy until an agreed-upon documented product definition is reached.

Many investigators consider initial screening, including market-related activities, idea generation including preliminary evaluation, and concept evaluation including technical-related activities as phases of the fuzzy front end [6-1], [6-2], [6-3]. For instance, the fuzzy front end could incorporate the formulation of product strategy, identification of opportunity, idea generation, product definition, project planning, and management reviews [6-12]. To stress the need for integrating the fuzzy front end with realization phases, it was clearly stated by Khurana and Rosenthal [6-13], "The failure to integrate a product strategy, a well-planned portfolio, and a facilitating organization structure with clearly identified customer needs, a well-defined product concept, and a project plan can severely hamper New Product Development (NPD)."

To structure the NPD fuzzy front end, a range of models have been suggested [6-1], [6-8], [6-9], [6-12], and [6-13]. These models are structured for business, marketing, product, or technology context.

One of the systematic, transparent, and simply structured models is the "stage-gate model." The name of the "stage-gate model" comes from its modular structure with a continuation or termination gate after each stage, as shown in Figure 6.10 [6-1]. The stage-gate model represents a business context model considering both market and product dimensions. It also promotes organizational buy-in to an idea as it moves forward.

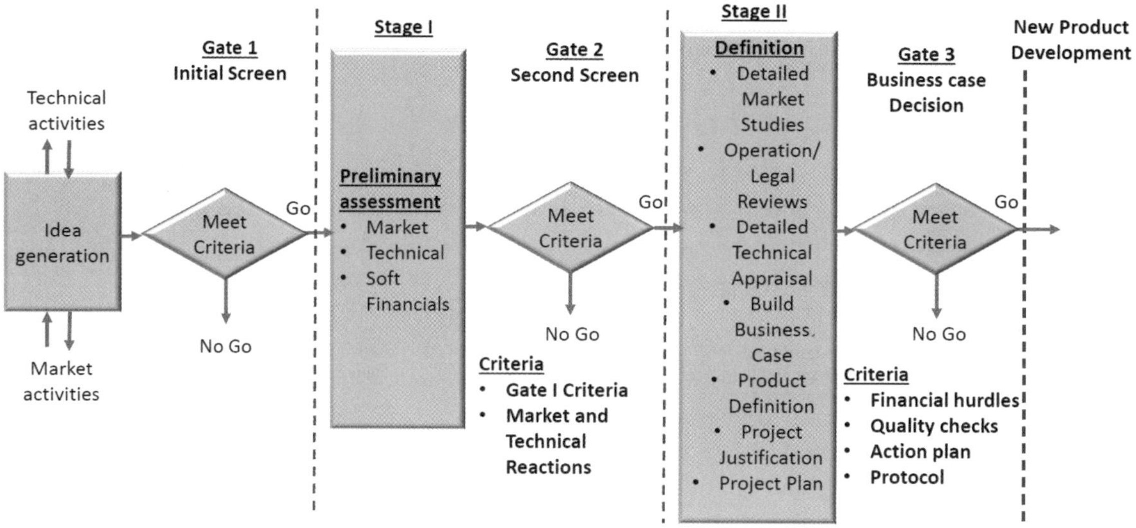

Figure 6.10 Basic structure of stage-gate model [6-1].

As shown in the Figure 6.10, the "stage-gate model" has three decision gates starting with the initial screening after idea generation and ending with the business case decision. The decision to continue or terminate the project at each gate is usually made by a cross-functional team.

Despite its simplicity, transparency, and logical structure, the stage-gate model lacks flexibility. Furthermore, some creative product ideas may be terminated prematurely before supplementary development. However, using the stage-gate model to structure the front end could be more effective with incremental innovations than having no structure [6-20].

Two general approaches for linking front-end activities were recognized from surveys conducted by Khurana and Rosenthal [6-12]. The first approach, mostly adopted by U.S. firms, "relies on a formal process to lend some order and predictability to the front end" [6-9]. The second approach is prevalent among the Japanese firms where "companies strive to foster a company-wide culture in which the key participants in front-end activities always remain focused on the following considerations: business vision, technical feasibility, customer focus, schedule, resources, and coordination" [6-12]. It was also suggested from the surveys that "the front-end approach must be compatible with the firm's product, market, and organizational contexts. For example, standardized approaches seem to work best for incremental innovations [6-12]." As a result of their research and based on empirical evidence, Khurana and Rosenthal developed a process model that integrates project-relevant input with foundation elements, as shown in Figure 6.11 [6-12]. The project-relevant input includes product ideas, market analysis, and technology options. The foundation elements include product strategy, portfolio plans, organizational culture, and capabilities.

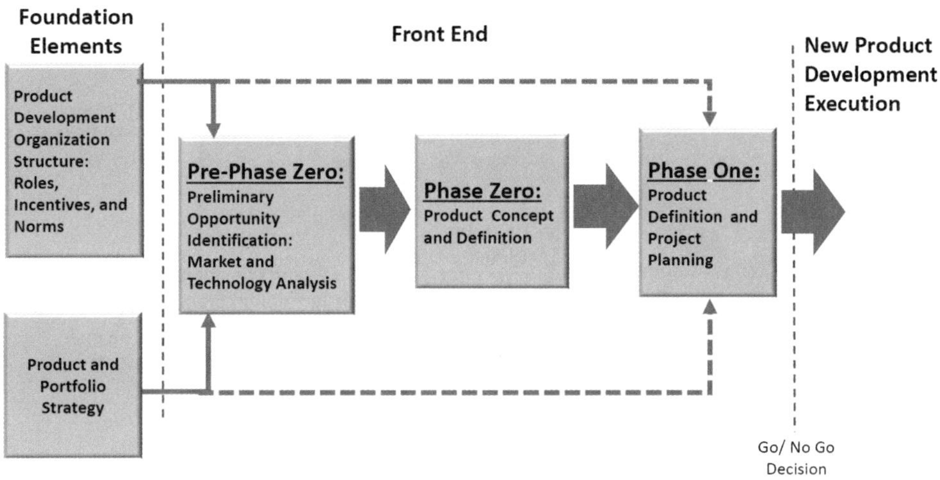

Figure 6.11 Khurana and Rosenthal front-end model [6-12].
(Reprinted with permission of MIT Sloan Management Review, www.sloanreview.mit.edu).

As shown in Figure 6.11, the prephase zero begins when an opportunity is identified. Phase zero is initiated if the opportunity is worth pursuing by assigning a team to work on the product concept and definition. Finally, the front-end activities are concluded by the business decision for accepting or rejecting the product definition and the project plan developed by the team at the end of phase one. While the Khurana and Rosenthal model [6-12] links business strategy, portfolio strategy, and product definitions, it does not detail the activities to be followed [6-21].

Several other front-end approaches and models have been developed. One of these models is the learning-based approach. In this approach, the organizational transformation needed for major innovation arises from continuous learning through trial and errors [6-22], [6-23]. Other nonserially structured front-end innovation models have been suggested. One of such models, shown in Figure 6.12, is the new concept development (NCD) model created by Koen et al. [6-24].

The NCD model consists of three major elements: the engine, the influence factors, and the five controllable activities. The engine in the core represents the organizational internal drives, leadership, and culture. The influencing factors on the boundary represent the internal and external constraints such as resource limitations and government regulations. The five controllable activities represent opportunity identification, opportunity analysis, idea generation and enrichment, idea selection, and concept definition [6-24]. NCD projects begin with opportunity identification or idea generation, as indicated by the green arrows. The larger yellow arrow represents how concepts leave the model and enter product development or another technology stage-gate process.

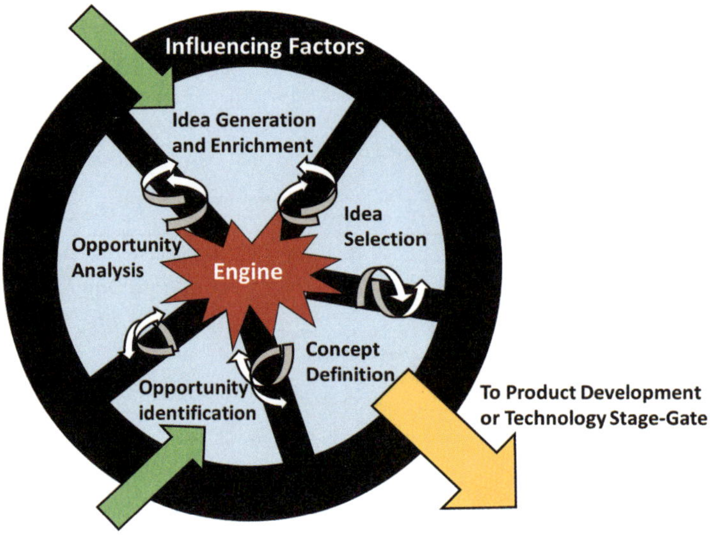

Figure 6.12 The NCD Koen et al.'s model [6-24].
(Copyright 2016 from "Providing Clarity and a Common Language to the Fuzzy Front End" by Peter Koen et al. Reproduced with permission of the Industrial Research Institute, http://www.iriweb.org.)

Building on the Khurana and Rosenthal model [6-12] with the addition of opportunity analysis and identification [6-24], [6-25], [6-26], through environmental screening for potential business opportunities, Hüsig et al. [6-27] presented the process model shown in Figure 6.13. In this model, there are three different phases and three gates: the first phase is environmental screening, the second phase is idea generation, and the third phase concept definition and business planning. Front-end activities are concluded when a go or no-go decision is made. Unlike the stage-gate process, the activities of the three phases are not in sequential order. Instead, the front-end team is usually working on several ideas and projects in parallel by relaying ideas and concepts between phases [6-27]. With a deeper look, this serially structured model embodies parallel controlled activities similar to the NCD model. Aside from adding the needed gates, project, and business planning activities, the model combines opportunity identification with environmental screening and integrates idea selection with idea generation activities.

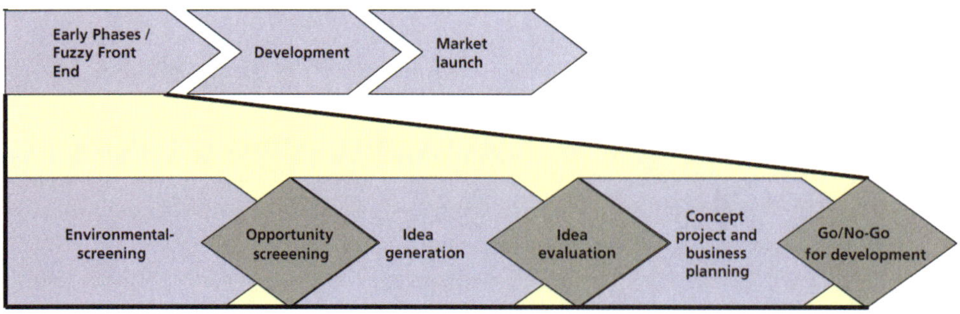

Figure 6.13 The structure of Hüsig et al.'s model [6-27].

6.5 Preinception Model for Front-End Innovation

To clearly separate the holistic innovative pursuit from the structured sequential product realization activities, a preinception model for the front-end innovation is needed. This model should include many advantages of the front-end innovation models discussed in the previous section. It should also address the main two conflicting issues in product realization: the need for an unrestricted flow of creative pursuits and the planned and timed deliverables of the product realization commitments. The model should seamlessly integrate with most ongoing organizational activities and fit naturally with a typical organizational structure.

For a front-end innovation model to be an integral part of the ongoing organizational activities, three groups of concurrent, continuous, and interconnected preinception activities should be conducted in conjunction with the central and core activity of opportunity identification. These three main activities can be identified as the organizational pursuit, environmental screening, and idea generation. Also, like any of the realization phases, having identifiable subphases can greatly streamline the preinception phase, outline the activities, and regulate needed inputs. These identifiable subphases are the idea kernel inception, idea kernel conception, and idea kernel maturation, shown in Figure 6.14.

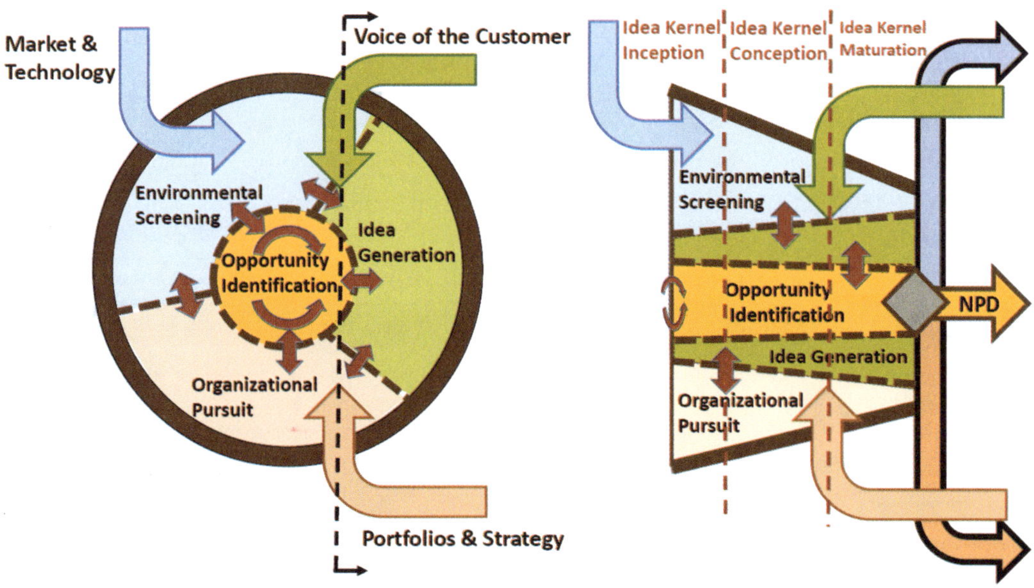

Figure 6.14 The preinception front-end innovation model.

At the end of the front end of the innovation funnel, the kernel of the preincepted idea is evaluated against a set of predefined criteria. Based on the evaluation outcome, as shown in Figure 6.15, the preincepted idea kernel is considered as one of the three main categories of NPD (inception), product improvement (conception or maturation phase), or technology improvement (R&D technology conception or maturation activities).

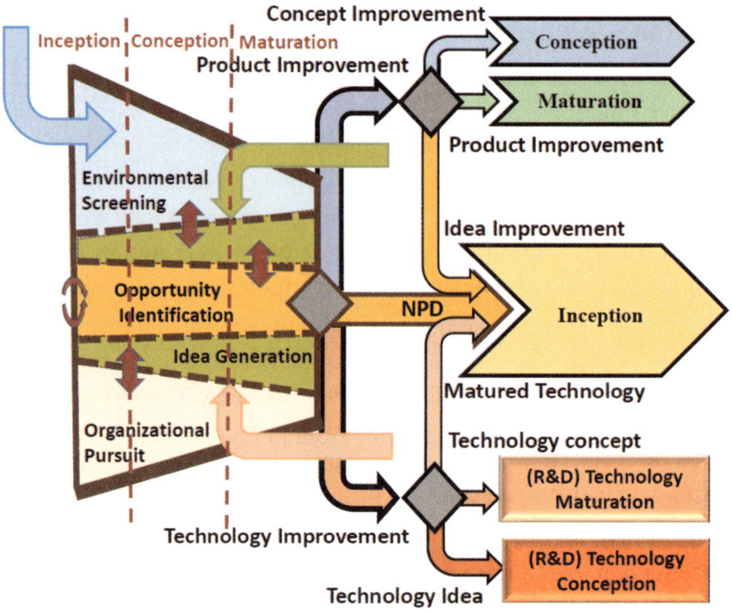

Figure 6.15 Different product and technology improvement levels in the preinception front-end model.

With the integration of the front end, the full product realization process, including the preinception phase, can be modeled as shown in Figure 6.16.

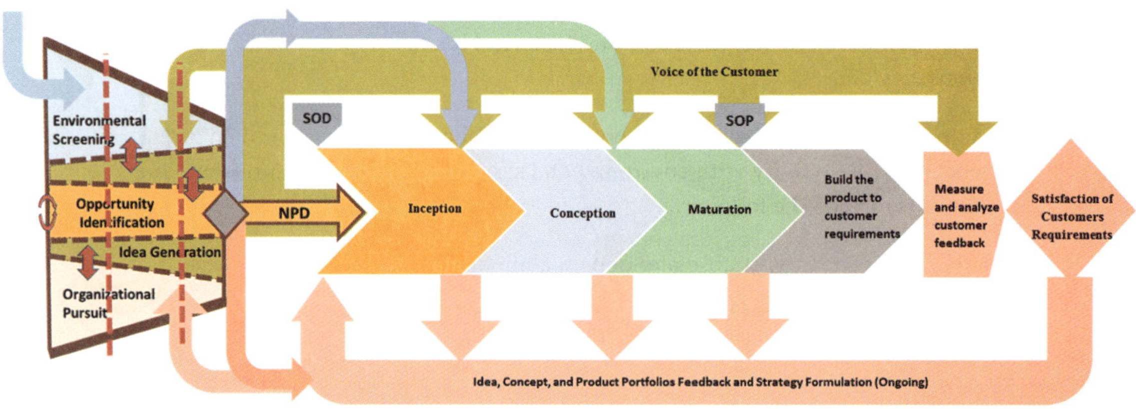

Figure 6.16 The full product realization process including the preinception front end.

In general, the activities in each of the front-end groups depend on the organizational culture, structure, processes, and capabilities. The following discussion for each front-end group outlines some of the activities that could be included.

6.5.1 Idea Generation

Idea generation is the beginning of ideation. The idea generation group of activities includes gathering, conceiving, and formulating the kernels of the new ideas. In addition to idea genesis and maturation as the core activities, the group shares with environmental screening the VOC activity for complete understanding and alignment with customer's perceptions and aspirations. The group shares with the organizational pursuit group the information from the idea, concept, product, and technology portfolios. The main outputs of idea generation group activities are:

- *VOC:* Preferences, priorities, and expectations
- *Idea Kernel:* Preinception list of product key features, characteristics, and attributes
- *Product/Process/Technology Assumptions:* From portfolios' experiences

6.5.2 Organizational Pursuit

Organizational pursuit is a front-end group that coordinates and integrates the inputs of several ongoing organizational activities and resources. These activities embody the organizational capability of several departments or sub-organizations, such as the following:

- strategic planning
- R&D
- business/marketing innovations and benchmarking
- idea, concept, product, and technology portfolios
- customers' experience, feedback, and historical quality/dependability and warranty data

In addition to sharing all the information from portfolios with idea generation group and benchmarking activity with environmental screening, the organizational pursuit main outputs include the following:

- historical warranty/quality and dependability data
- technology/product reliability studies
- business/marketing strategy

6.5.3 Environmental Screening

Environmental screening is the organizational group of activities continuously monitoring customers, competitive products, market, and technology. This group of activities encompasses customer, market, and technology current state and future trends including the following:

- market information, segmentation, and trend
- market modeling and simulation
- technology screening, information, and trend
- customer surveys, information, and trend
- strength, weakness, opportunity, and threat analyses

In addition to sharing VOC data with idea generation and benchmarking with organizational pursuit, the environmental screening group key outputs from the ongoing scanning, screening, and analyses include the following:

- market research
- technology report
- product/process benchmark data

6.5.4 Opportunity Identification

The opportunity identification core activity provides the continuous drive, direction, and connectivity to all other preinception activities. In addition to sharing and coordinating with all other activities, the main contribution of the opportunity identification group is the execution team experience.

Figure 6.17 illustrates the different activities of each front-end group. While information sharing is continuous between the main groups and between the subgroups, the three key shared activities (titled in red) in addition to opportunity identification are as follows:

Figure 6.17 Different front-end activities.

- VOC activities (shared between idea generation and environmental screening)
- Portfolio activities (shared between idea generation and organizational pursuit)
- Benchmarking activities (shared between organizational pursuit and environmental screening)

Figure 6.18 illustrates the main output for each activity. These outputs, identified by the activity color, will be discussed in more detail in the next chapter.

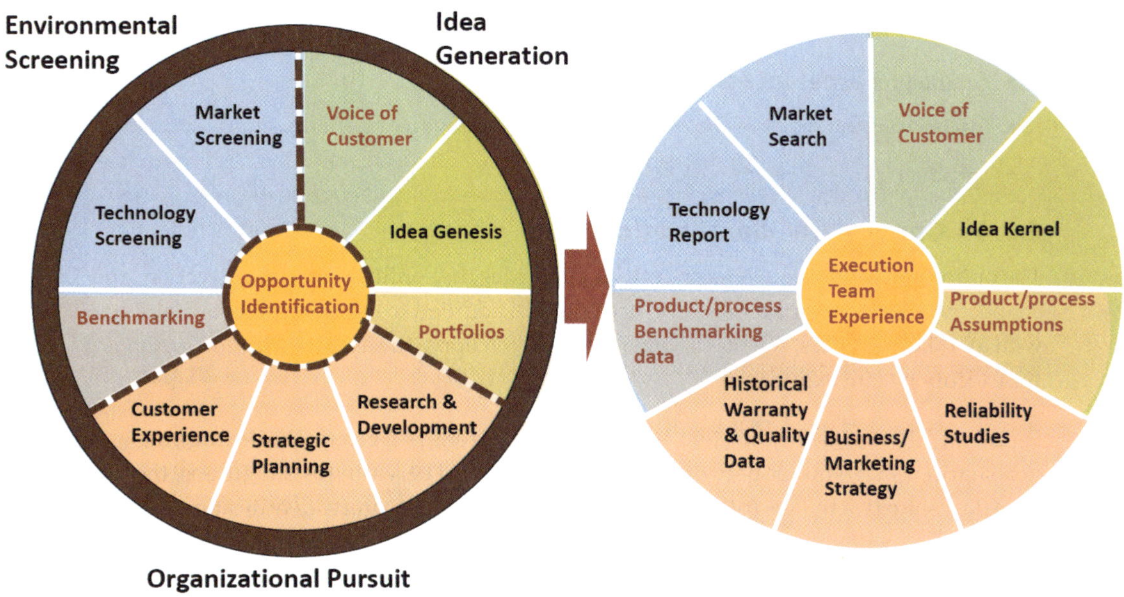

Figure 6.18 Main outputs for each front-end activity.

These outputs in addition to other needed information constitute the input to the perceptual realization phase of NPD (or New Product Realization NPR). In summary, these outputs include the following:

- idea kernel
- VOC
- market research
- technology report
- product/process benchmark data
- historical warranty/quality data
- business/marketing strategy
- reliability studies
- product/process assumptions
- execution team experience

Depending on the type of findings and required inputs, the continuous exchange between the four front-end groups can be initiated at any group and proceed to the other. For example, environmental screening could identify several potential opportunities through communication with the organizational pursuit. These potential opportunities could then be communicated to an opportunity identification group where one or few opportunities are identified and directed to idea generation activities. Similarly, an identified opportunity or idea could trigger environmental screening and so on. Once an idea is identified, the inception, conception, and maturation subphases identified in Figure 6.14 are followed by the idea generation group, in connection with the other groups, for the idea kernel realization.

6.5.5 Idea Kernel Inception

Idea inception is the result of the ongoing front-end activities of organizational pursuit, environmental screening, opportunity identification, and idea generation. A recognized potential opportunity or idea by any activity can be shared and circulated until a potential idea is identified and directed to the idea generation group. The identified potential idea is enriched and specified during the inception phase by the idea generation activity in connection with opportunity identification and inputs from the organizational pursuit and environmental screening. Inception is completed when the potential idea is enriched, specified, and transformed into an idea kernel.

6.5.6 Idea Kernel Conception

Idea kernel conception fledges the incepted idea kernel into the idea concept of the desired product form, function, and fit. Accordingly, desired product features, characteristics, and attributes are shaped during idea kernel conception. Conception is completed by populating the desired list of product key characteristics and attributes.

6.5.7 Idea Kernel Maturation

Maturation prepares the Idea kernel concept for the structured inception phase through development and validation. The Idea kernel concept features, characteristics, and attributes are matured through identification of production process, technology, internal, and external constraints. Several iterations of development and validation with inputs from other front-end groups may be needed during maturation. Maturation is completed when the developed idea kernel is validated and all necessary product realization issues including process, technology, internal and external constraints are considered.

Starting with the idea kernel inception, the external scanning and organization pursuit inputs are tapered down, as indicated by the funnel shape shown in Figure 6.14. Meanwhile, the idea generation activities are slightly ramped-up until the idea kernel maturation is completed.

The main decision gate at the end of the front end of innovation funnel is shown in Figure 6.15. At this decision gate, the matured idea kernel is evaluated against a set of pre-established criteria for consideration as one of the three main categories of NPD, product improvement, or technology improvement.

A cross-functional team from front-end and product development activities must determine if the idea kernel is ready for NPD. If the idea does not meet the criteria to merit NPD, then the idea could be implemented on a current product as product improvement or it may need further technology improvement. Whether the idea is product improvement or technology improvement, the level of improvement must be decided for proper routing. In general, there are three identifiable levels of product improvements:

- product improvement (to be implemented on current product maturation)
- concept improvement (to be implemented on current product conception)
- idea improvement (to be implemented on current or future product inception)

Similarly, three identifiable levels of technology improvement are as follows:

- matured technology (for technology implementation on current or future product inception)
- technology concept (for technology maturation by R&D activities)
- technology idea (for technology conception by R&D activities)

As shown in Figure 6.15, at the product improvement decision gate, product improvement and concept improvement idea kernels are directed to the appropriate phase for implementation on a current or future product improvement cycle. Idea kernels having the potential for integration with current or future idea inception are directed to NPD. At technology improvement decision gate, idea kernels with the need for further technology improvement are directed to the appropriate research and development (R&D) activity. Idea kernels with available matured technology are directed for implementation with NPD.

To avoid unnecessary risks on product development process outcomes and timing, premature implementation of unvalidated technologies for NPD or product improvements should be avoided. To achieve initial quality, and long-term dependability targets only matured and validated technologies should be considered. In other words, the introduction of unvalidated technology, or the development of new technologies during product realization, may have an adverse effect on the realization process outcomes and timing in addition to causing product quality and dependability issues. The following examples illustrate successful implementation and introduction of technologies with different levels of innovations:

6.5.7.1 Example 6.1: Toyota Prius Implementation of Matured and Validated Technologies: To improve fuel economy, Toyota announced in November 2014 that the next-generation 2015 Toyota Prius hybrid will utilize road-tested microchips and microcontrollers of the Toyota's TS040 Hybrid Le Mans racer, shown in Figure 6.19 [6-28].

Figure 6.19 Toyota's TS040 hybrid Le Mans racer.
(Reprinted with permission from TOYOTA MOTOR SALES, U.S.A., INC.)

For radical innovations, the new technologies could be introduced to advanced concept vehicles for maturation and validation. Until matured and validated, implementation of these technologies in mainstream product realization may prove to be risky. To demonstrate the implementation of new and radically innovative technologies in advanced concepts, the following example is considered.

6.5.7.2 Example 6.2: GM EN-V Advanced Concept Vehicles for Radical Product and Technology Innovations: To radically improve driving safety, efficiency, and independence, GM introduced the electric networked autonomous driving vehicle concept Chevrolet EN-V, shown in Figure 6.20. The Chevrolet EN-V does not use gasoline, can avoid crash, and can drive and park itself [6-29].

The preinception front-end innovation model discussed could be easily integrated with most organizational structures since many of the front-end groups and activities are already established. To succeed, however, these activities should be aligned with organizational strategies, aspirations, vision, mission, and goals. In addition, the individuals involved with the front-end activities should have vast product realization experience and the ability to think creatively and holistically.

Figure 6.20 GM's EN-V autonomous driving concept, 2014 [6-29]. (Reprinted with permission from GM.)

6.6 Learning Outcomes from Front End of Innovation

The desired learning outcomes from studying the front end of innovation in vehicle realization can be categorized as follows.

6.6.1 Knowledge

Innovation is a context-dependent endeavor with varying dimensions and levels. In the context of technology development, innovation could mean improved technique or new application. In the context of product development, innovation could mean new or improved form and/or function. In the context of marketing, innovation could mean market forecast and/or market strategy innovations. In the context of organizational business, innovations could include marketing and/or technology innovations. As for innovation levels, three main levels or types could be identified as incremental innovations, major innovations, and radical innovations. The uncertainty and risks associated with innovations are usually proportional to the innovation level. For example, higher uncertainties and risks are associated with major innovations than incremental innovations.

In the context of the auto industry, it is important to make the distinction between the product innovation levels and the innovation levels of the technologies employed in the product. It is also necessary to make the distinction between innovations and improvements in products and technologies. For example, not all vehicle and technology improvements are results of innovations, and not all innovations cause improvements. Also, some of the vehicle and technology innovations risks may be passed on to the customers. These passed on risks, however, are the main cause for losing customers' loyalty. Therefore, the key to successful vehicle and technology innovations is to minimize the associated customers' risks.

New product realization front-end activities are mostly perceptual. Perceptual realizations are in general less crisp than virtual realizations, which in turn are less crisp than physical realizations. Accordingly, front-end activities such as screening including market-related activities, idea generation including preliminary evaluation, and concept evaluation including technical-related activities are labeled the fuzzy front end. To structure the NPD fuzzy front end, a range of models have been suggested. To separate the holistic innovative pursuit from the structured sequential product development activities, a preinception model for front-end innovation is proposed. This model combines several advantages of the previous fuzzy front-end innovation models. It addresses the main two conflicting issues in product development: the need for an unrestricted flow of creative pursuits, and the structured planned and timed deliverables of costly product development commitments. While the proposed model could be easily integrated with most organizational structures, the front-end activities must be aligned with organizational strategies, aspirations, vision, mission, and goals.

6.6.2 Skills

Building the skills for identifying the context, dimensions, and levels of innovation, along with the associated uncertainties and risk, is necessary. Also, developing the ability to distinguish between the product innovation levels and the innovation levels of the technologies employed in the product along with their potential impact is essential. Ultimately, vast product realization experience, teamworking skills, and creative holistic thinking are needed for successful participation in the front end of innovation activities.

6.6.3 Paradigm/Attitude

While there are tremendous growth opportunities in innovating and taking risks, product and technology innovation risks should never be passed on to the customers. In the long term, incremental vehicle innovations with minimum customers' risks are much more effective and conducive to organizational growth than radical innovations with higher customers' risks.

6.7 References

6-1. Cooper, R. "Stage-Gate Systems: A New Tool for Managing New Products," *Business Horizons*, May–June, 1990.

6-2. Koen, P., Ajamian, G., and Burkart, R. et al. "Providing Clarity and a Common Language to the Fuzzy Front End," *Research Technology Management*. 44(2):46–55, 2001.

6-3. Kim, J., and Wilemon, D. "Focusing the Fuzzy Front End in New Product Development," *R&D Management*. 32(4):269–279, 2002.

6-4. Herstatt, C., Verworn, B., and Nagahira, A. "Reducing Project Related Uncertainty in the 'Fuzzy Front End' of Innovation: A Comparison of German and Japanese Product Innovation Projects," *International Journal of Product Development*. 1(1):43–65, 2004.

6-5. Herstatt, C., Stockstrom, C., and Verworn, B., et al. "Fuzzy Front End Practices in Innovating Japanese Companies," *International Journal of Innovation and Technology Management*. 3(1):43–60, 2006.

6-6. Verworn, B., Herstatt, C., and Nagahira, A. "The Impact of the Fuzzy Front End on New Product Development Success in Japanese NPD Projects." Working Papers/ Technologie- und Innovations Management: Technische Universität Hamburg-Harburg, 39, 2006.

6-7. Smith, P., and Reinertsen, D. 1991. Developing Products in Half the Time. Van Nostrand Reinhold: New York.

6-8. Murphy, S., and Kumar, V. "The Front End of New Product Development: A Canadian Survey," *R&D Management*. 27(1):5–16, 1997.

6-9. Verworn, Birgit, and Cornelius Herstatt. "Approaches to the 'Fuzzy Front End' of Innovation." (2). Working Papers/Technologie-und Innovationsmanagement: Technische Universität Hamburg-Harburg, 1999.

6-10. Nagahira, Akio, Isao Sugiyai, Cornelius Herstatt, Birgit Verworn, C. Stockstrom, et al. "Impact Analysis of Front End Practices in Innovative New Product Development in Japanese Manufacturing Companies." *Technology Management for the Global Future, 2006. PICMET* 6:2586–2594. 2006.

6-11. Frishammar, Johan, Henrik Florén, and Joakim Wincent. "Beyond Managing Uncertainty: Insights From Studying Equivocality in the Fuzzy Front End of Product and Process Innovation Projects." *IEEE Transactions on Engineering Management*. 58(3): 551–563, 2011.

6-12. Khurana A., and Rosenthal, S. "Integrating the Fuzzy Front End of New Product Development," Sloan Management Review, winter, 1997.

6-13. Khurana A., and Rosenthal, S., "Toward Holistic 'Front Ends' in New Product Development," *Journal of Product Innovation Management*. 15(1):57–74, 1998.

6-14. Lynn, G., and Akgun. A., "Innovation Strategies Under Uncertainty: A Contingency Approach for New Product Development," *Engineering Management Journal*. 10(3): 11–17, 1998.

6-15. http://www.toyota-global.com/company/history_of_toyota/75years/data/automotive_business/production/production/overview/index.html

6-16. http://www.sae.org/images/magazines/135/autofeb02.jpg.

6-17. http://www.sae.org/congress/2002/wed.pdf.

6-18. http://money.cnn.com/2002/01/08/autos/auto_tech/.

6-19. http://www.digitaltrends.com/cars/gm-and-honda-announce-hydrogen-fuel-cell-collaboration-can-they-bring-hydrogen-to-the-mainstream/.

6-20. Cooper R. "Overhauling the New Product Process," *Industrial Marketing Management*. 256: 465–482, 1996.

6-21. Backman, M., Börjesson, S., and Setterberg, S. "Working with Concepts in the Fuzzy Front End: Exploring the Context for Innovation for Different Types of Concepts at Volvo Cars," *R and D Management*. *37*(1):17–28, 2007.

6-22. Lynn, G., and Green, C. "Market Forecasting for High-Tech vs. Low-Tech Industrial Products," *Engineering Management Journal*. 10:(1): 15–18, 1998.

6-23. Frishammar, Johan, Henrik Florén, and Joakim Wincent. "Beyond Managing Uncertainty: Insights from Studying Equivocality in the Fuzzy Front End of Product and Process Innovation Projects." *IEEE Transactions on Engineering Managemen*t. 58(3): 551–563, 2011.

6-24. Koen, P. A., G. Ajamian, R. Burkart, A. Clamen, J. Davidson, R. D'Amoe, C, et al. Wagner. "New Concept Development Model: Providing Clarity and a Common Language to the 'Fuzzy Front End'." *Research Technology Management* 44(2):46–55, 2001 doi 10.1080/08956308.2001.11671418.

6-25. Nobelius, Dennis, and Trygg, Lars. "Exploring the 'Fuzzy Front End' of the Product Development Process," *Proceedings of the 5th International New Product Development Management Conference*, Como, 777–792, May 25–26, 1998.

6-26. Verworn, Birgit: "The 'Fuzzy Front End' of Product Development: An Explorative Study," *Proceedings of the 9th International Product Development Management Conference*, Sophia Antipolis, France, 2002.

6-27. Hüsig, Stefan, Stefan Kohn, and Jarno Poskela. "The Role of Process Formalisation in the early Phases of the Innovation Process." *Proceedings of the 12th International Product Development Management Conference*, Copenhagen. 2005.

6-28. http://www.autonews.com/article/20141109/OEM06/311109995/toyota-to-pack-track-tested-tech-into-next-prius.

6-29. http://media.gm.com/media/us/en/gm/news.detail.html/content/Pages/news/us/en/2011/Oct/1012_Chev_ENV.html..html.

Chapter 7
Vehicle Perceptual Realization

"The customer's perception is your reality."

—Kate Zabriskie

7.1 Introduction

For any new product, the preinception phase is usually followed by perceptual realization as part of the organizational activities for product and technology innovations. Accordingly, perceptual realization signals the beginning of the fully structured new product realization. As discussed in Chapter 2, realization subphases are repeats of the three realization phases of inception, conception, and maturation in a fractal-like pattern. Therefore, product idea inception is the first subphase of the perceptual realization phase of any new product. This first perceptual realization subphase is then followed by the second subphase of product idea conception. Finally, the perceptual realization phase is concluded by the third subphase of product idea maturation.

In this chapter, the perceptual realization phase for new product development is discussed with two main objectives. The first objective is to illustrate the steps of maturing the preincepted idea kernel into a developed, validated, and planned product idea, or perceptually realized product concept. The second objective is to demonstrate the alignment of the realization phases with the main gates and milestones of the APQP [7-1].

7.2 Perceptual Realization in New Product Development

With front-end integration, perceptual realization or product inception is the first structured phase in new product development, as illustrated in Figure 7.1.

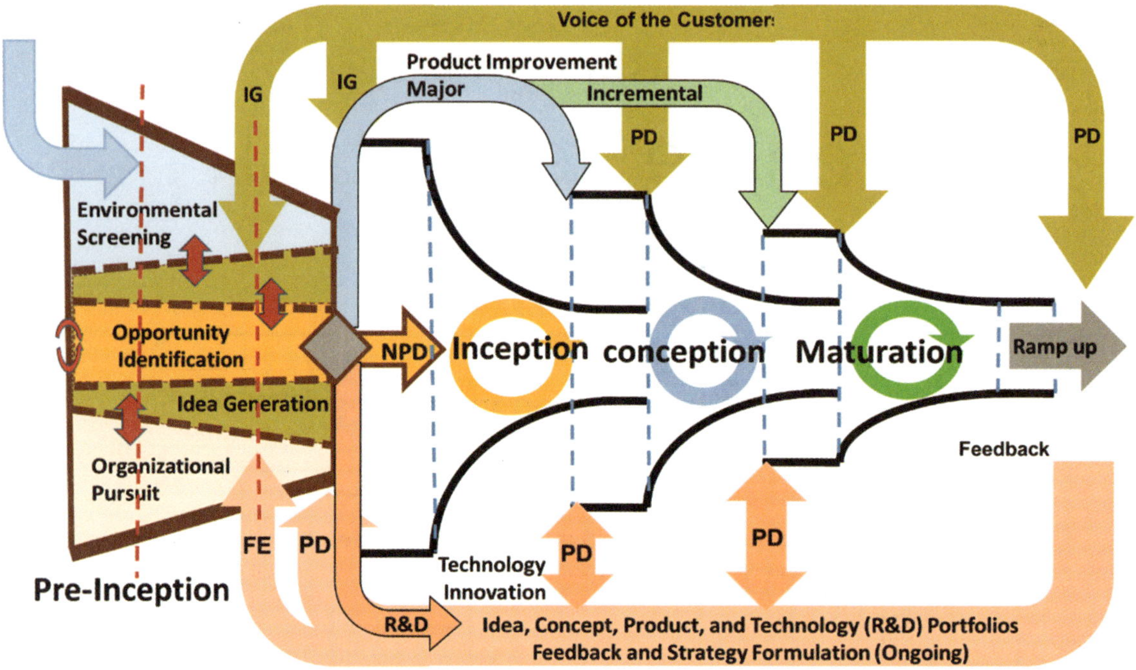

Figure 7.1 The full new product realization process.

While the preinception activities are continuous nonsequential, perceptual realization activities are the beginning of the structured, planned, timed, and sequential full product realization process. Figure 7.1 shows the full new product realization processes including the preinception, product improvement, and technology innovation routes. Following Figure 7.1, the subphases of the product inception or perceptual realization phase can be illustrated, as shown in Figure 7.2.

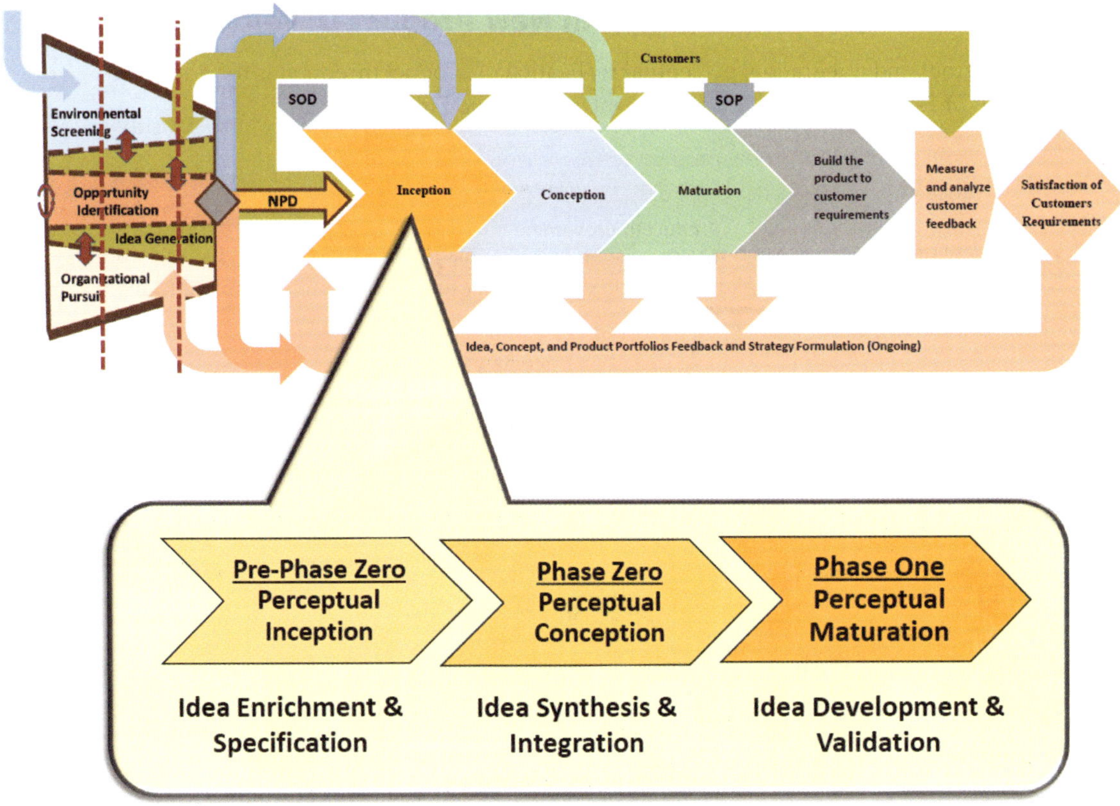

Figure 7.2 Perceptual realization is the first phase of structured product realization process.

As shown in Figure 7.2, the first subphase in perceptual realization is product idea inception or idea enrichment and specification. If the need for new product development became obvious during the idea kernel conception of preinception, then the inception subphase of perceptual realization could be integrated with the preinception maturation phase. Therefore, as an interface, and to some extent, a continuation to preinception, this perceptual realization subphase can be considered as prephase zero in new product development.

The second subphase in perceptual realization is product idea conception or idea synthesis and integration. Transformation of the product idea to a perceptual product concept begins during this subphase. Therefore, the perceptual conception subphase can be considered as phase zero in new product development.

Finally, the third subphase in perceptual realization is product idea maturation or idea development and validation. During this subphase, the perceptual product concept is matured through several development and validation iterations. Accordingly, from a product development perspective, this perceptual realization subphase can be considered as phase one of the new product development.

Fully developed and validated product with no quality gaps is the desired outcome of any new product realization process. For achieving desired product quality, the plan-do-study-act product quality planning cycle shown in Figure 7.3 [7-1] can be applied to the full product development cycle.

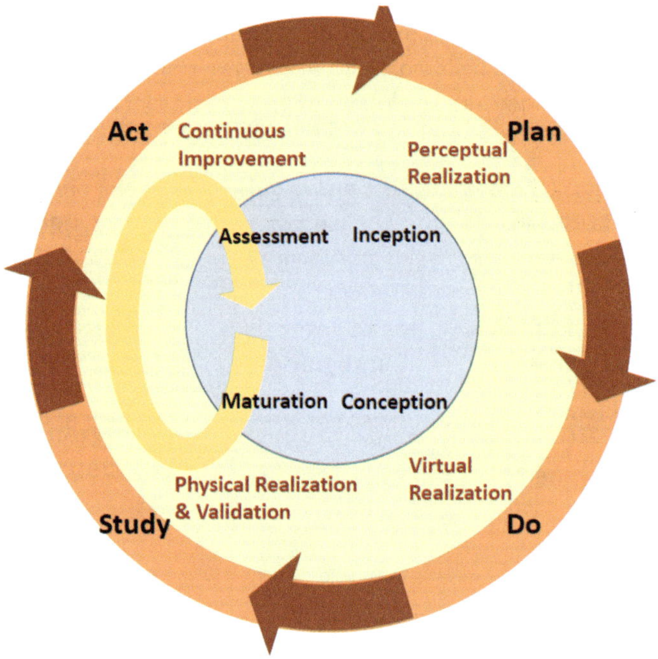

Figure 7.3 Full product development quality planning cycle.

During the new product realization cycle, both product and production process should be realized simultaneously from inception to maturation. As indicated in Figure 7.3, iterations between maturation and assessment are continued until all necessary corrections and improvements are validated and implemented for the product and production process.

Due the repeated pattern or the fractal sense discussed earlier, the plan-do-study-act product quality planning cycle can also be applied to every phase of the realization process. During perceptual realization, the product idea is specified, integrated, developed, and validated as close as possible to the voice of the customer (VOC). Therefore, this product idea must be closely aligned with the customer's perceptual realization of the desired product with no or minimum negative gaps as possible. As discussed in Chapter 5, these are the gaps between producer's imaginative perceptual realization and customer's imaginative realization of desired product characteristics and attributes contained in the array G_p. These negative gaps could exist due to missing customers' expected and desired characteristics and attributes or due to the presence of unexpected and undesired characteristics and attributes. As the reaction of the customer can be exponential with proportional exponential losses to the producer, closing the negative gaps and increasing the positive is a major step toward a quality product that attracts and excites customers.

Starting with the perceptual inception phase, the product quality planning cycle can be applied to each phase for closing or minimizing negative gaps. During perceptual maturation, iterations should be continued between perceptual development/validation and corrections/improvements until all negative gaps in the array G_p are closed or minimized. Accordingly, the perceptual realization structure from inception to maturation will be well aligned with the product quality cycle, as shown in Figure 7.4.

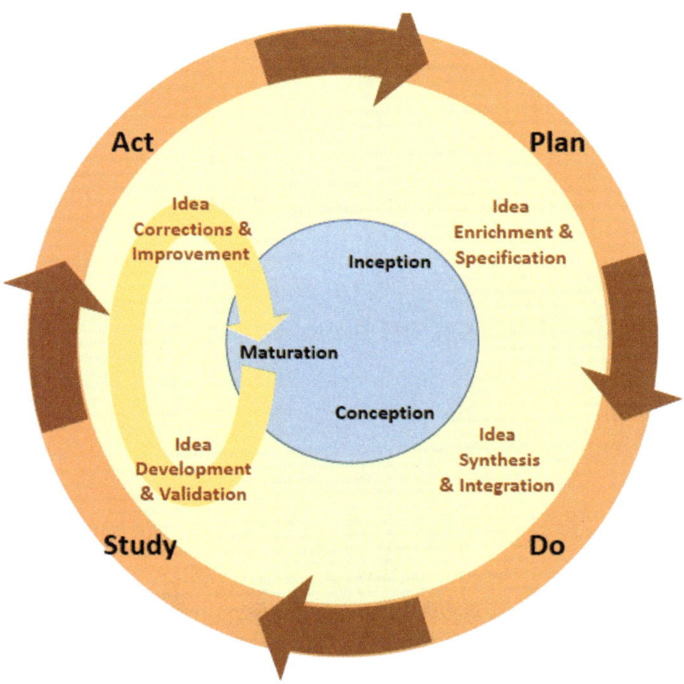

Figure 7.4 Perceptual realization quality planning cycle.

During perceptual realization, product and production process should be perceptually realized simultaneously from perceptual inception to perceptual maturation. Therefore, as integral parts of the same idea, both product and process should be perceptually realized jointly as outlined in Figure 7.5. Iterations during perceptual maturation should be continued until all necessary development and validation corrections and improvements are implemented for both product and production process.

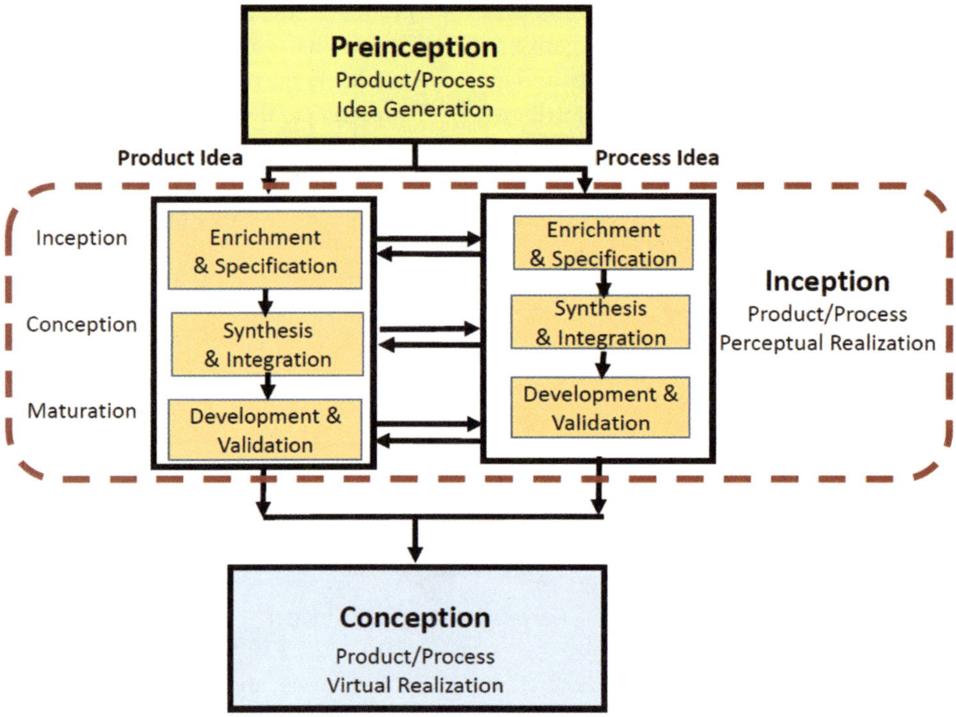

Figure 7.5 Joint perceptual realization of product and production process.

7.3 Perceptual Realization Model

For new product development, perceptual realization is the start of structured, timed, and sequential realization activities. Building on the Khurana and Rosenthal model [7-2], [7-3], the structured perceptual realization activities for the product and production process idea can be outlined, as shown in Figure 7.6.

As shown in Figure 7.6, the structured product perceptual realization model connects the preinception outputs to perceptual product idea inception, conception, and maturation. The prephase zero begins with the preinception outputs to further enrich and specify the idea. Starting with the specified idea, phase zero transforms the idea into a perceptual product concept through further synthesis and integration of the product idea genes. Perceptual realization is concluded by the maturation of the perceptual concept definition of the product, production process, and project plan.

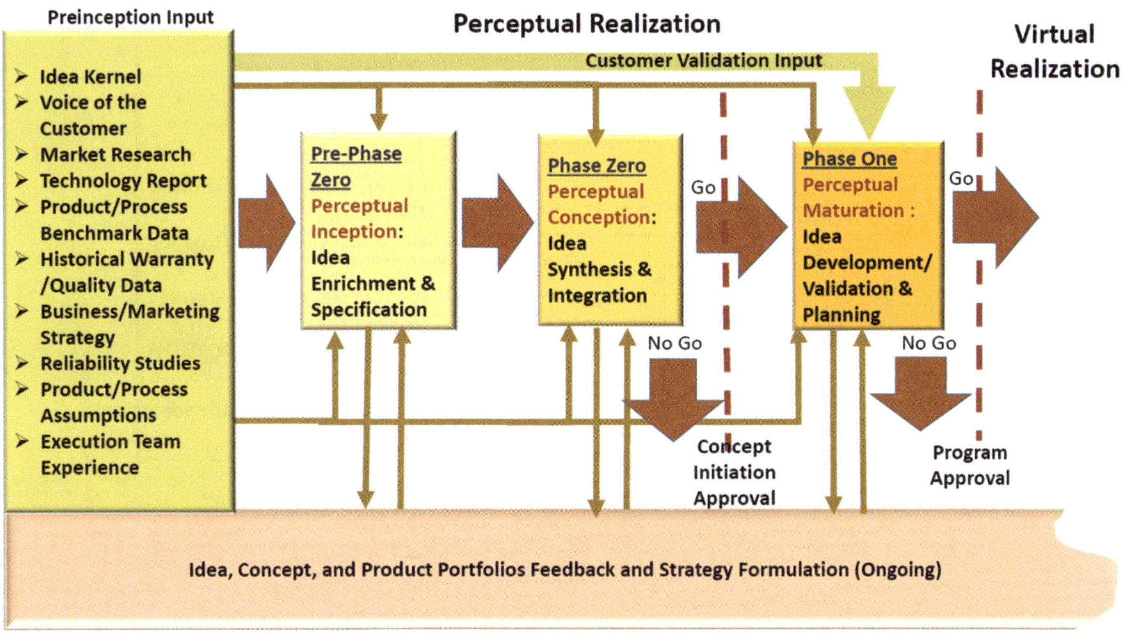

Figure 7.6 Perceptual realization model.

Iterations inside and between perceptual inception, conception, and maturation activities are possible if needed. However, iterations are more likely to be needed during the maturation phase between development/validation and corrections/improvements, as shown in Figure 7.4, until all perceptual realization issues and improvements are completed. During conception, any needed specification adjustments or changes could be made without going back to inception. During maturation, any required adjustments or changes resulting from validation reviews can be addressed through further development, without having to go back to conception. Customers and business reviews are conducted during the development and validation iterations until a decision is made to proceed to virtual realization or send the perceptual concept to idea portfolio. In the following, the key inputs and outputs in addition to the inception, conception, and maturation activities of the perceptual realization model, shown in Figure 7.6, will be discussed.

7.4 Perceptual Realization Inputs

For new product realization the inputs are the preinception outputs shown in Figure 7.7.

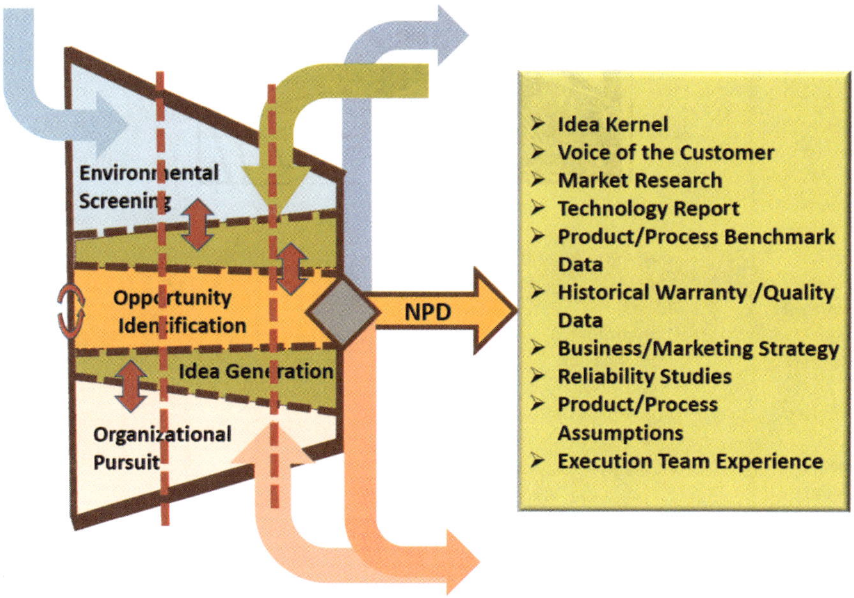

Figure 7.7 Inputs to perceptual realization of new product development.

Each input from the preinception front-end activities, shown in Figure 7.7, can be summarized as follows.

7.4.1 Idea Kernel

The idea kernel is the key output of the preinception and the main input to perceptual realization structured activities. As the culmination of the idea generation activities during preinception, the idea kernel contains the synthesis of all other inputs and activities as it pertains to the generated idea goals. In addition to the ideation sketches for interior as shown in Figure 7.8 [7-4], and exterior, as shown in Figure 7.9 [7-5], these goals include the following:

- development goals and objectives
- preliminary list of desired product key characteristics and attributes
- preliminary list of aspired targets for key characteristics and attributes

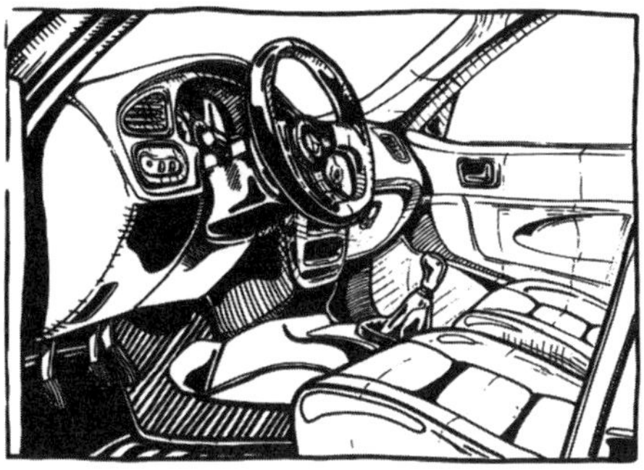

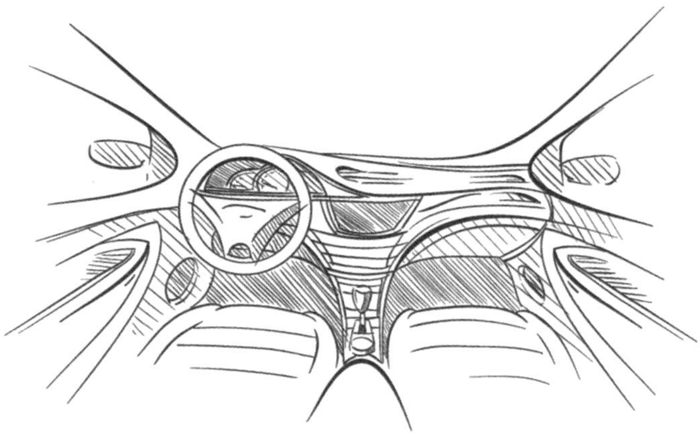

Figure 7.8 Interior ideation sketches [7-4].
(Babich Alexander (top)/Roman Ya (bottom)/shutterstock.com)

Figure 7.9 Exterior ideation sketches [7-5]. (Roman Ya/shutterstock.com)

7.4.2 Voice of the Customer

Product form, function, and global characteristics are ultimately evaluated in the customer's cognitive, affective, and psychomotor domains. From the organizational perspective, it is vital to have the ability to recognize, understand, and eventually predict responses of these customer's domains. The most direct access to these customer's domains is the VOC.

As ongoing preinception and product development activities, the VOC should continuously depict the customer cognitive domain with focus on the imaginative and receptive realizations. The main goal of the cognitive VOC input is to provide the accurate portrayal needed to minimize the gaps between the organizational and customer perceptual realizations of the potential new or improved product. Equally important are the customer's affective domain responses, expressed by the emotional reactions associated with met or unmet expectations, priorities, and preferences. These affective VOC inputs can improve the organizational ability to forecast customer's responses to product gaps and predict associated quality issues. Finally, the VOC activities should continuously observe the customer's psychomotor tendencies and behaviors. These observable psychomotor behaviors can be directly used in shaping potential product architectural and utility characteristics and attributes.

Currently, there are many methods to solicit the VOC [7-6]. Many of these methods employ customer visits, customer comments, customer advisory groups, and warranty cards. As the VOC activities are continuous from preinception to product development [7-7], it is useful to employ the most effective methods for the product development phase being executed. For example, there are VOC methods that are more suited for idea generation (IG input in Figure 7.1), such as the following:

- ethnographic research (observing customers for extended periods)
- customer visit teams (comprehensive interviews)
- customer brainstorming (for new ideas)
- community of enthusiasts (organizationally nurtured for new ideas)

There are also VOC methods that are more suited for product development (PD input in Figure 7.1), such as the following:

- customer focus groups (to validate ideas, concepts, and products)
- customer advisory boards or panels (to guide product improvement and new development)
- customer or user designs (product design input through internet and crowdsourcing)
- open innovation (using partnerships and supply chain for new off-the-shelf technologies)

7.4.3 Market Research

Market research is ongoing front-end activities used to identify and study the state of the market and to forecast market trends. To guide early phases of product development, communicated market research outputs may include the following:

- market research findings, analysis, and rationale
- market current state and trend
- market forecast and future trends
- targeted market segments
- targeted customer and need

While the product's key characteristics, attributes, and targets are selected based on the VOC, the targeted markets and customers are selected based on thorough market research. Ideally, it is the recognized need during market research that motivates the opportunity identification, and propels the new product development or technology innovation. Once the need is recognized the targeted customer is accordingly identified.

In case targeted customers are identified before the need, then the needs of these customers must be studied for opportunity identification. This could be the case for loyal and return customers. Loyal customers should in fact be the base upon which an organizational business is founded. Therefore, understanding and serving the need of loyal customers should be the top priority of organizational pursuits. To ensure business success and organizational stability, loyal customers must be satisfied before venturing to attract new customers. To further understand customers' loyalty in the auto industry, the following case study is considered.

7.4.3.1 Case Study 7.1: Customers' Loyalty Analysis: The published results of J. D. Power dependability studies for the two years 2013 and 2014 show that about 56% to 57% of the "owners who reported no problems stayed with the same brand" [7-8]. However, "brand loyalty slipped" to about 41% to 42% "among owners who reported three or more problems" [7-8].

From these results, it can be observed that there are three emerging categories of customers in market trend:

- about 41% to 42% return customers independent of the experience
- about 14% to 16% nonreturn customers due to unsatisfactory experience
- about 43% to 44% noncommitted customers independent of the experience

Depending on organizational strategy, organizational pursuit could be focused on taking care of the 41% to 42% return customers. At the same time, assuring that customer satisfaction has the potential of retaining an additional 14% to 16% of customers. Finally, understanding and serving the needs of noncommitted customers has the potential of retaining and attracting new customers.

7.4.4 Technology Report

The VOC alone may not be sufficient to guide technology adoption or development. In general, the VOC provides the necessary inputs for identifying key product characteristics and attributes along with some desired targets and technologies. With limited technological expertise, the VOC is not include the best technology to use unless the specific technology being surveyed is made available. In any case, decisions must be made about both product desired characteristics and attributes and best technology to achieve them. Ideally, product features and performances should be dictated by the customer while the technology needed to achieve these demands are decided by the experts.

To stay in the forefront of technology and technological innovations, continuous external environmental screening for current and emerging technologies is essential. Successful opportunity identification for new technologies and technological innovations in product development or improvement depends on continuous external and internal screening. Without continuous screening of technology state and emerging trends, disruptive technologies could threaten organizational competitiveness and stability. To guide early phases of product development, technology reports may include:

- external technology state findings and analysis
- technology forecast and future trends
- internal technology state, readiness, and timing
- targeted internal technologies and their development/validation state
- targeted external technologies and partnerships

7.4.5 Product/Process Benchmark Data

Benchmarking is an essential ongoing activity for guiding organizational pursuit and opportunity identification with input from environmental screening. It is usually used to measure an organizational pursuit issue relative to the best-in-class. Benchmarking can be classified as formal or informal. Formal bench marking can also be classified as performance benchmarking or best practice benchmarking [7-9]. For the entire organization, benchmarking should include strategic, financial, performance, process, and product benchmarking. For the product, the focus should be on benchmarking of the product and production process characteristics, attributes, and targets. For these types of benchmarking, the key steps may include [7-9]:

- competitive product identification
- planning and preparation for the benchmarking activity
- measuring and collecting relevant data during benchmarking
- analysis and evaluation of results
- documentation and communication of findings and insights

7.4.6 Historical Warranty and Quality Data

Historical warranty and quality/dependability data provide vital and accurate validation measures of previous products and process capabilities. These data contain the incident reports of accumulated millions of miles of customers' usage and experiences. Therefore, these data are essential for product development, process improvement, and supply chain building. Accordingly, these data should be made available at all product realization phases and should become more specific as the development process progresses. Sources for this data include the following:

- warranty reports
- J. D. Power initial quality and dependability reports
- issue resolution reports
- dealerships' and customers' surveys
- plant and field returns

7.4.7 Business/Marketing Strategy

The identified opportunity and generated idea are shaped by the market research from environmental screening and the business strategy established by organizational pursuit. Therefore, the identified customers' needs and organizational rationale to meet these needs must become the mission for planning, technical undertakings, and subsequent decision making. The business and market strategy output may include the following:

- market need
- competitive products
- product positioning
- resource allocation
- target cost and timing

7.4.8 Reliability Studies

Product and technology reliability studies should contain the accumulated correlations between actual field failures and corresponding validation tests. These studies should include product architectural, performance, and health characteristics. It should also be part of the ongoing organizational pursuit activities. Reliability studies of new targeted technologies should also be conducted and reported. Results of these studies are necessary to establish proper targets and representative development/validations tests.

7.4.9 Product/Process Assumptions

During preinception activities, there are usually several assumptions made during different group activities and idea generation. These assumptions include product attributes, process capability, state of technology, and customers' input. Considering these assumptions during perceptual realization is essential to have a complete picture and understanding.

7.4.10 Execution Team Experience

A key factor in identifying the opportunity and synthesizing the idea during the critical embryonic phases is the human element. The knowledge, skills, attitudes, and paradigms of the preinception team greatly shape the idea generation outcomes. Some of these experiences that a team brings to the activities include the following:

- innovative and creative abilities
- integration and facilitation skills
- tacit knowledge
- previous product realization experiences
- things that went right and things that went wrong [7-1]

Documenting some of the team experiences can provide useful insights regarding the kernel formation phase. These documentations may include the following:

- thought process
- evolution history
- creative and innovative views associated with idea generation
- insights and experiences
- unresolved and foreseen issues

7.5 Perceptual Inception

Perceptual inception or prephase zero is the interface and, to some degree, an extension to the preinception front-end activities for the portion of ideation, as shown in Figure 7.10, of the full realization process.

As shown in Figure 7.10, the full new product realization process and the realization phases can be naturally aligned with the main gates and milestones of the APQP [7-1]. Since perceptual inception is the interface and, to some degree, an extension to the preinception front-end activities, the transition from preinception to perceptual inception can be gradual during prephase zero, as demonstrated in Figure 7.10. In any case, the outputs of the preinception discussed earlier are the inputs to the perceptual inception.

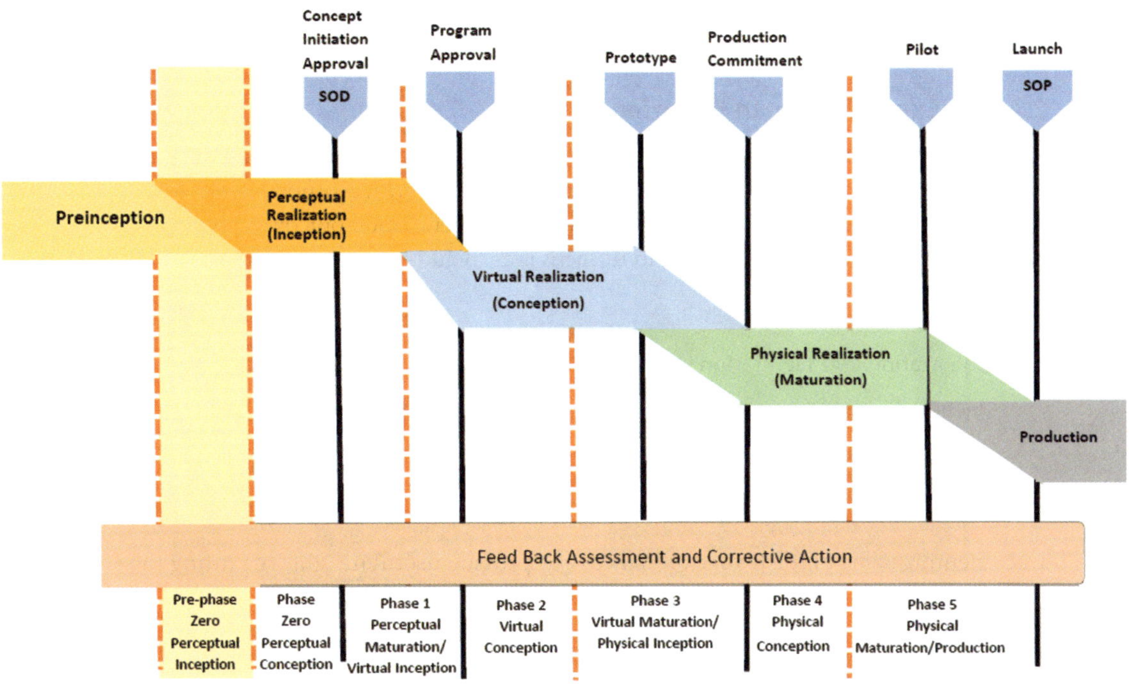

Figure 7.10 The perceptual inception phase of new product realization process.

Starting with the idea kernel and guided by the additional customers' inputs from the customers' reviews during preinception maturation and portfolios' experiences, the idea kernel can be enriched. The preliminary bill of product aspirations (bill of product) or the full matrix of aspired product characteristics, attributes, criteria, and targets is specified. Also, the preliminary bill of process aspirations (bill of process) or the full matrix of aspired process characteristics, attributes, criteria, and targets is initiated. The rest of the preinception inputs should be used to guide the product idea enrichment and specification, and then carried on to guide the perceptual conception at phase zero. Shaping and specifying interior and exterior styling are also completed during this phase, as shown by the examples of the interior sketch in Figure 7.11 [7-10], and exterior sketch in Figure 7.12 [7-11].

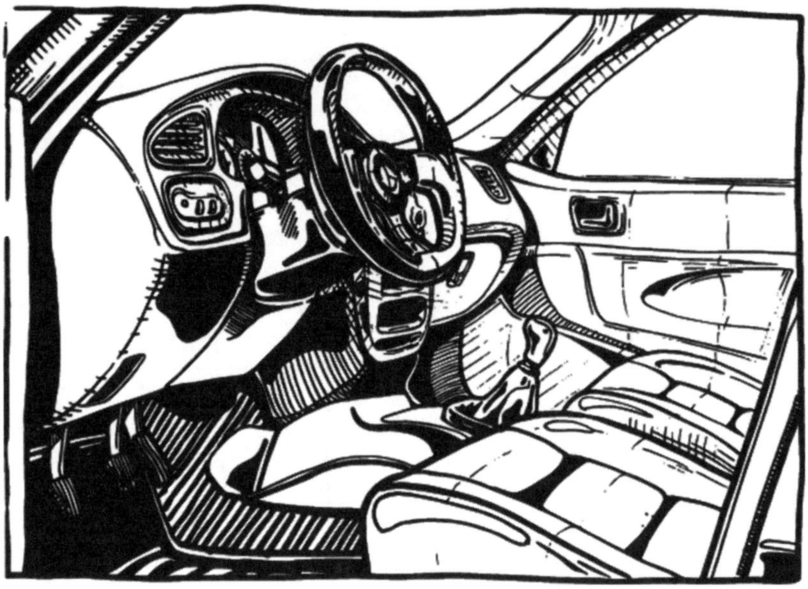

Figure 7.11 Specified interior sketch [7-10]. (Babich Alexander/shutterstock.com)

Figure 7.12 Specified exterior sketch [7-11]. (benedekbox/shutterstock.com)

The perceptual inception or idea enrichment and specification begins with developing the proper set of attributes for each characteristic. The necessary criteria for each attribute and targets for each criterion are specified as shown by the illustrative example in Table 7.1.

Table 7.1 Illustrative example of characteristics, attributes, criteria, and targets matrix (bill of product).

Characteristics	Attributes	Criteria	Targets
Safety	Crushability	Frontal impact	Five Stars
		Side impact	Five Stars
		:	:
Durability	Dynamic resistance	Fatigue loading	100,000 mi
		Input excitations	26-Hz body beaming frequency
			18-Hz body torsional frequency
		:	:

Using the house of quality, the bill of product aspirations or the full matrix of aspired product characteristics, attributes, criteria, and targets can be developed, as shown in Figure 7.13.

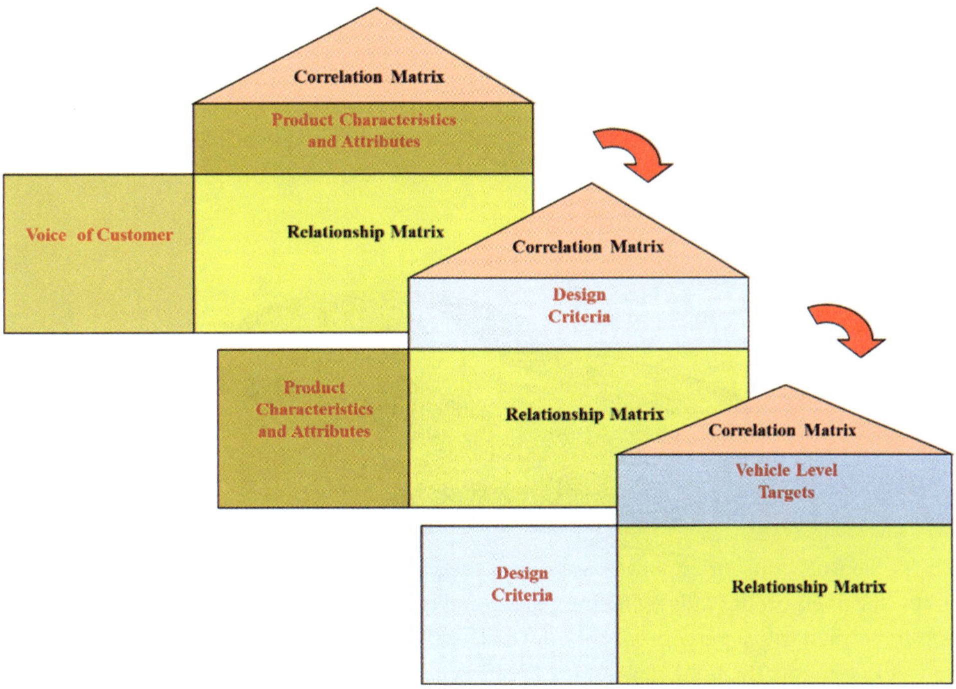

Figure 7.13 The house-of-quality sequence from the VOC to vehicle-level targets.

Consequently, the output list of perceptual inception, as shown in Figure 7.14, includes the following:

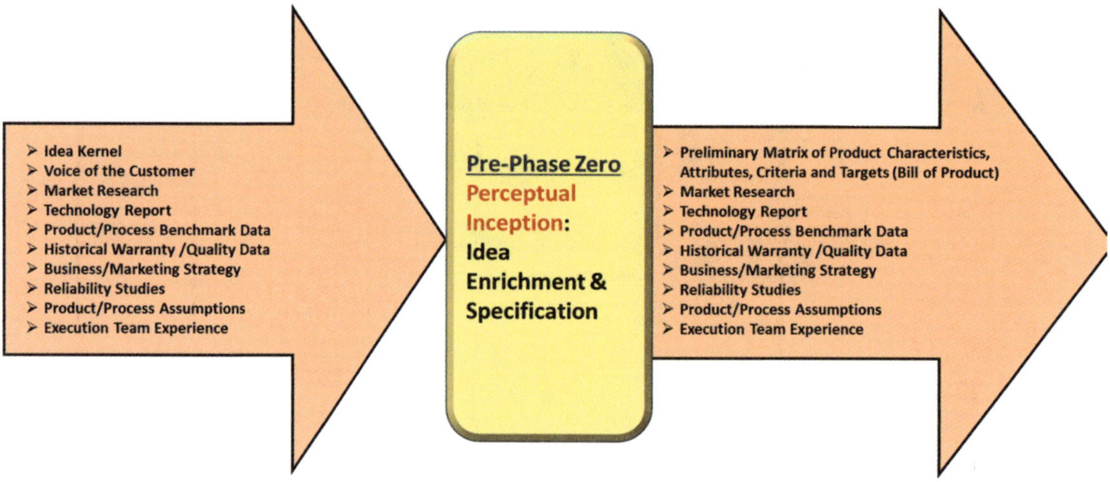

Figure 7.14 Perceptual inception inputs and outputs.

- preliminary matrix of product characteristics, attributes, criteria, and targets (bill of product)
- market research
- technology report
- product/process benchmark data
- historical warranty/quality data
- business/marketing strategy
- reliability studies
- product/process assumptions
- execution team experience

7.6 Perceptual Conception

The perceptual conception, or phase zero, is the perceptual synthesis and integration of the specified product idea. During perceptual conception, the incepted product idea is transformed into a perceptual product concept. Therefore, phase zero or perceptual conception marks the final stage of the ideation activities encompassing idea preinception, inception, and conception. Before starting phase one or perceptual maturation, concept initiation approval must be decided at the end of perceptual conception. Typically, concept initiation approval marks the traditional start of development (SOD), as shown in Figure 7.15.

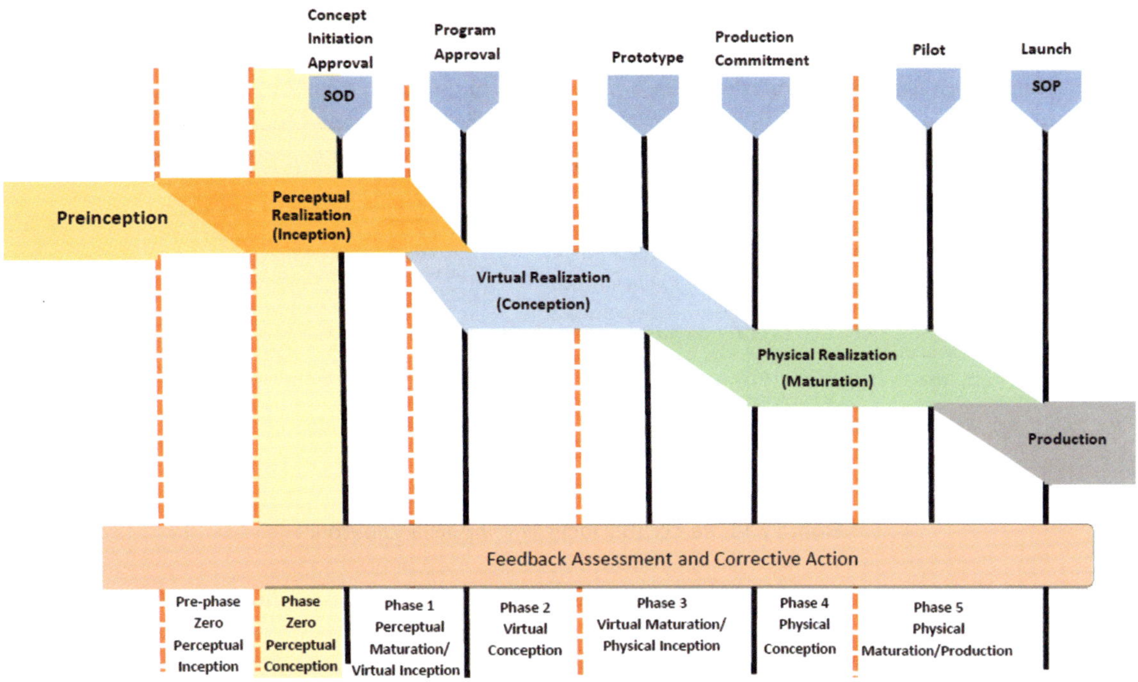

Figure 7.15 The perceptual conception phase of new product realization process.

Perceptual architectural and performance integrations are the main activities during perceptual conception. Selecting the key subsystems and components to meet form, function, and fit targets to populate the preliminary bill of materials (BOMs) is the main objective for these activities. Guided by the benchmarking studies and process assumption, the aspired characteristics, attributes, criteria, targets, and process flowchart of both product and production process can be initiated and shaped. Accordingly, the product desired characteristics, attributes, criteria, and targets and the process desired characteristics, attributes, criteria, and targets are further formulated as the BOMs are being populated.

Vehicle concept architectural packaging, conflicting targets, integration, and interface issues must be addressed during perceptual conception phase. Therefore, some initial basic vehicle-level architectural and performance integration assessment studies must be performed to address these key issues. Addressing and resolving these issues during perceptual conception is vital for the success of the subsequent phases of the vehicle realization process. To perform these studies, previous experiences, expert opinions, in addition to simplified modeling and simulation virtual tools and/or physical mockups, can be utilized. If the key integration issues are not resolved, then the idea conception should be halted during the concept initiation approval review gate, shown in Figure 7.9, until resolutions are achieved or the perceptual concept is recycled.

Using the correlation matrix in each of the house-of-quality sequence shown in Figure 7.13, early detection of potential conflicts can be identified. In addition to the must-have performance attributes, the main issues that must be satisfactorily resolved during perceptual conception are the architectural, performance, and manufacturability ones, particularly exterior and interior styling and packaging. For example, some of the styling conflicts and issues include the following:

- conflicts between styling concept and packaging of identified subsystems and components
- conflicts between styling concept and safety requirements
- conflicts between styling concept aerodynamics and fuel economy targets
- conflicts between styling concept and durability, maintainability, and reparability targets
- conflicts between styling concept and manufacturability requirements

It is always easier and less costly to eliminate existing or potential conflicts and create synergy between attributes and targets through creative alternatives at the perceptual conception phase. During perceptual conception, there is always more freedom to identify options that will cause conflicting targets to be decoupled [7-12], [7-13].

By creatively resolving any existing or potential conflict during perceptual conception, the exterior and interior styling concepts can be finalized, as shown by the examples of the interior sketch in Figure 7.16 [7-14] and the exterior sketch in Figure 7.17 [7-15].

Figure 7.16 Interior perceptual concept sketch [7-14].
(Babich Alexander/shutterstock.com)

Figure 7.17 Exterior perceptual concept sketch [7-15]. (benedekbox/shutterstock.com)

In addition to the finalized exterior and interior styling concepts, the output list of perceptual conception, as shown in Figure 7.18, includes the following:

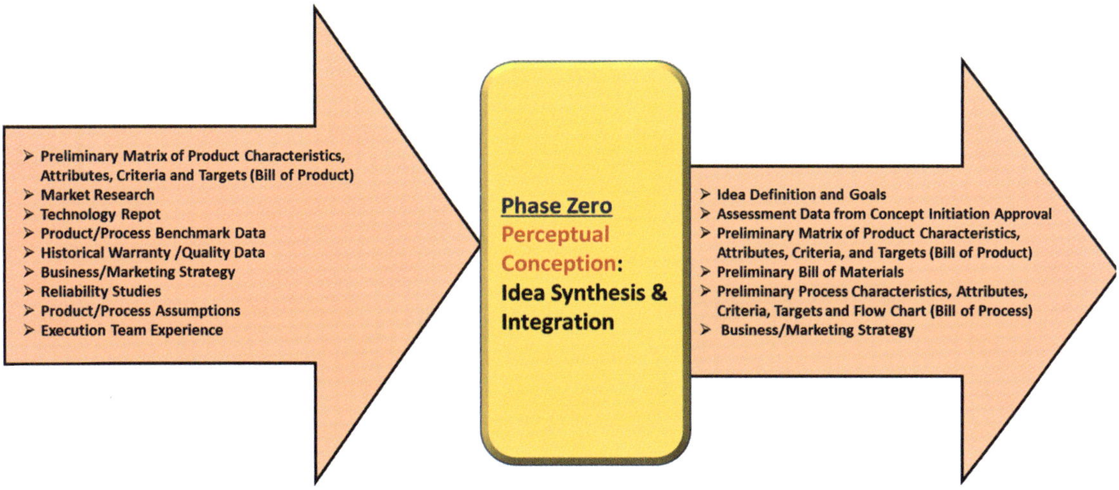

Figure 7.18 Perceptual conception inputs and outputs.

- idea definition and goals
- assessment data from concept initiation approval
- preliminary matrix of product characteristics, attributes, criteria, and targets (bill of product)

- preliminary bill of materials
- preliminary process characteristics, attributes, criteria, targets and flowchart (bill of process)
- business/marketing strategy

7.7 Perceptual Maturation

Perceptual maturation is the last phase of perceptual realization, as shown in Figure 7.19. Development and validation of the perceptual concept are the main activities during perceptual maturation or phase one. During the realization process and every realization phase, maturation prepares and transitions the product and production process to the next realization domain. As a result, a maturation phase can be easily augmented with the specification phase of the next domain. Accordingly, perceptual maturation can be augmented with the specification phase of virtual realization.

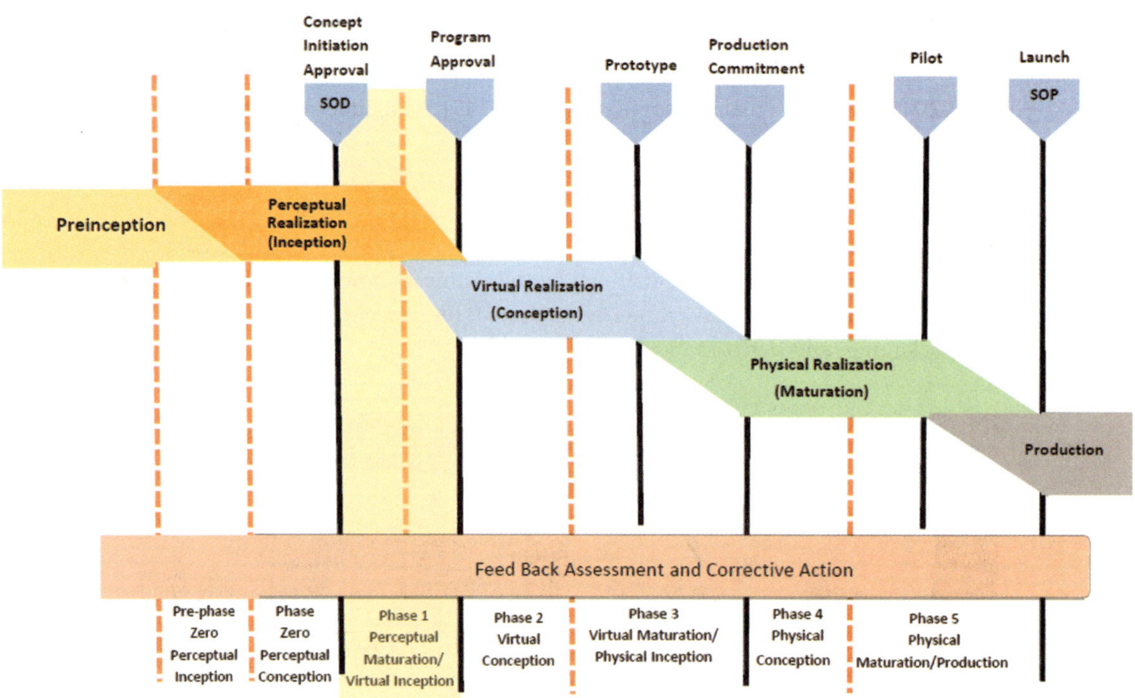

Figure 7.19 The perceptual maturation phase of new product realization process.

While virtual representation of perceptual concept elements could start earlier to facilitate communication, full representation of the perceptual concepts virtually can speed up the development and validation iterations. An example of rendered virtual interior and exterior of a vehicle perceptual concept is shown Figure 7.20 [7-16].

Figure 7.20 Rendered virtual interior and exterior [7-16]. (videodoctor/shutterstock.com)

With the rapid development and improvement of virtualization tools [7-17], creation and transformation of perceptual concept sketches into virtual representation is becoming the normal practice for aiding and enhancing perceptual realization. Virtual reality modeling and simulations, as demonstrated by the example shown in Figure 7.21 [7-18], as well as augmented reality, can immensely improve the perceptual maturation of styling, packaging, and performance.

Figure 7.21 Virtual reality modeling and simulations [7-18]. (Reprinted with permission from Ford.)

To illustrate the use of augmented reality as a virtual validation tool for perceptual conception, the following two case studies are considered.

7.7.1 Case Study 7.1: Augmented Reality Perceptual Validation Simulation

For perceptual concept maturation of in-wheel motor for small electric vehicles, an augmented reality simulation system was developed. The system was developed using a cyberglove and a motion capture system. Figure 7.22 shows an example of the augmented reality simulations [7-19].

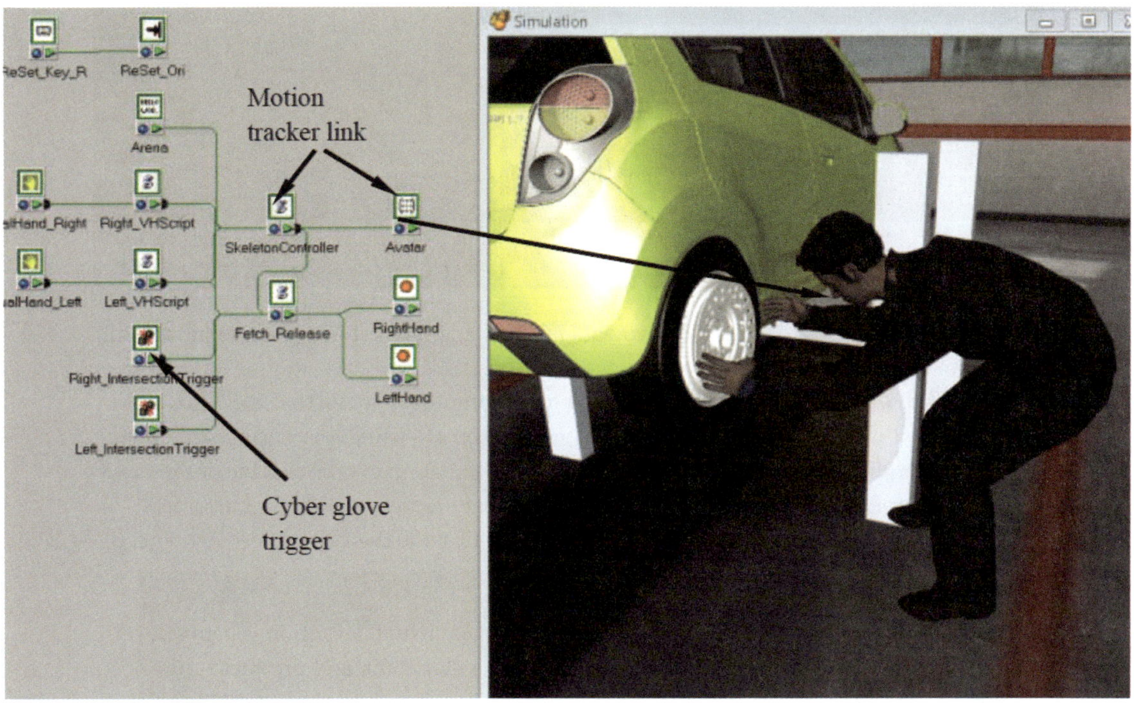

Figure 7.22 Perceptual concept maturation using augmented reality simulation [7-19].

7.7.2 Case Study 7.2: Immersive Visualization

The Ford Motor Co. Immersive Virtual Review (IVR) lab uses computer software and motion-tracking system to create a full-scale virtual vehicle representation. The IVR lab is used to evaluate future production vehicles and replace the 3-D physical bucks. As shown in Figure 7.23 [7-20], the lab simulates the look and feel of the vehicle environment allowing the occupant to evaluate many human-machine interactions, such as blind spots and reflections. The IVR system also allows for quick changes of various elements inside the lab's Programmable Vehicle Model (PVM) [7-17].

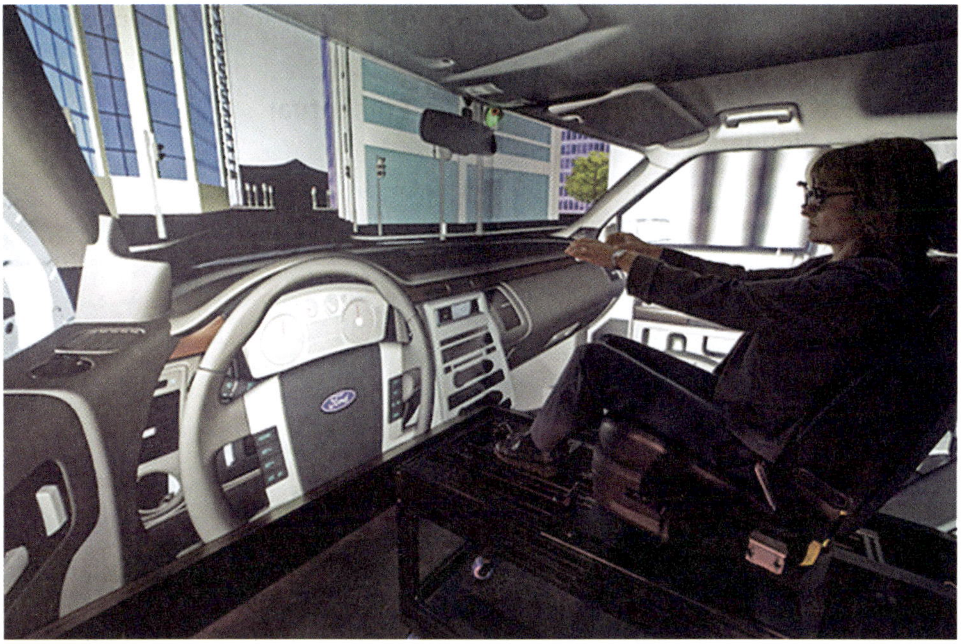

Figure 7.23 Immersive visualization system [7-20]. (Reprinted with permission from Ford.)

In addition to providing a common platform for communication, virtual tools can also facilitate the identification and accelerate the resolution of styling, packaging, and integration issues during perceptual development and validation iterations. One of the major impacts of these tools is providing an interactive visual representation for customers' review and feedback. Used correctly, these tools can provide the key to close the perceptual gap between customers and developers.

During perceptual maturation iterations, as the vehicle virtual definition is being developed and validated, project management, product development, and product assurance plan are developed. Virtual definitions and simplified simulations should validate that packaging, integration, and conflicting targets' issues are resolved for subsystems and components on the preliminary BOMs. Different alternatives and creative resolutions should be considered to complete the concept definition and ensure meeting customers' desired characteristics, attributes, and targets.

By the end of perceptual realization, a final validation customers' review should be performed. Vehicle virtual definition can facilitate the review and enhance visualization. Management review is performed after the final validation and customers' review are completed. Management approval and support must be secured to start virtual realization; otherwise, the perceptual concept is sent to the idea portfolio.

In addition to the virtual development and validation of exterior and interior perceptual styling concepts, the output list of perceptual maturation, as shown in Figure 7.24, includes the following:

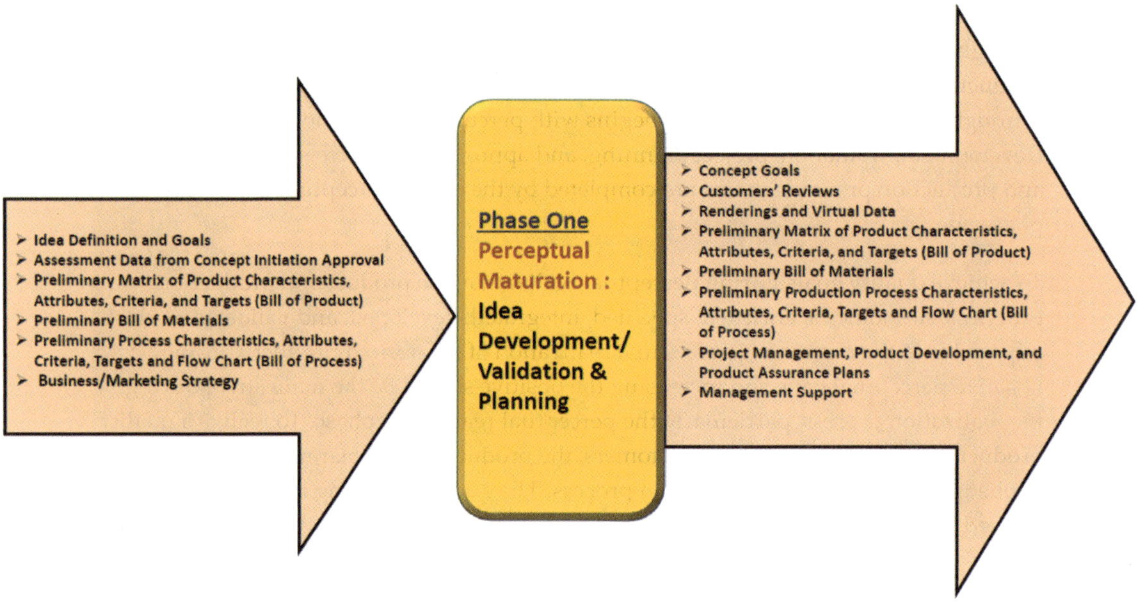

Figure 7.24 Perceptual maturation inputs and outputs.

- concept goals
- customers' reviews
- renderings and virtual data
- preliminary matrix of product characteristics, attributes, criteria, and targets (bill of product)
- preliminary bill of materials
- preliminary production process characteristics, attributes, criteria, targets, and flowchart (bill of process)
- project management, product development, and product assurance plans
- management support

7.8 Learning Outcomes from Vehicle Perceptual Realization

The desired learning outcomes from studying vehicle perceptual realization can be categorized as follows.

7.8.1 Knowledge

For new product development, perceptual realization is the beginning of the fully structured, sequential, and timed product realization process. Similar to any realization phase, identifiable perceptual realization subphases are perceptual inception, perceptual conception, and perceptual maturation. During perceptual inception or prephase zero, the preincepted idea kernel is transformed into product and production process idea through further enrichment and specification. The transformation of the product and production process idea into the perceptual product and production process concept (through synthesis and integration) begins with perceptual conception or phase zero. Development, validation, project planning, and approval of the perceptual product and production process concept are completed by the end of perceptual maturation or phase one.

To achieve quality goals during perceptual realization, the product and production process idea should be enriched, specified, integrated, developed, and validated as close as possible to the customers' perceptual realization of the desired product. Closing negative perceptual gaps and increasing the positive should be the main strive during the realization process, particularly the perceptual realization phase. To realize a quality product that attracts and excites customers, the product quality planning cycle is usually applied to the full product realization process. The quality planning cycle can also be applied to each realization phase, starting with perceptual realization.

Conflicting targets, vehicle architectural packaging, integration, manufacturability, and interface issues must be addressed during perceptual conception and validated during perceptual maturation. In addition to being necessary for the success of the subsequent realization phases, finding creative resolutions to product and production process issues during perceptual realization is more effective and less costly.

In addition to validating perceptual realization, virtual simulations can facilitate the identification and accelerate the resolution of conflicting characteristics, attributes, and targets. For example, virtual visualization modeling and simulation tools, such as virtual reality and augmented reality, can greatly improve perceptual realization and facilitate styling, packaging, integration, and manufacturability during perceptual conception and maturation. In addition to providing a common platform for communication, these tools can also help in closing the perceptual gaps between customers and developers.

7.8.2 Skills

As an early realization phase, perceptual realization requires honed technical and interpersonal skills. In addition to the technical skills and expertise necessary to perform assigned tasks, high level of holistic, creative, and integrative thinking abilities and skills are needed during all perceptual realization subphases. Along with these abilities and skills, teamworking skills are crucial for integrating the perceptions of many experts into a collective realization that meets and exceeds customers' expectations. Due to the perceptual nature of the product idea, idea enrichment, specification, conception, and maturation, precise and clear communication between all team members becomes a must-have. Accordingly, every team member of the group dealing with perceptual realization must have a commanding level of holistic thinking and communication skills.

7.8.3 Paradigm/Attitude

During product realization, the realized outcome is aligned with the nature of the realization domain. This paradigm is necessary for success in dealing with the challenges and taking advantage of the realization domain nature. For example, the early phases of new product realization are usually perceptual in nature. Among the many challenges of working in the perceptual domain are the fuzziness of perceptions, the fluid nature of perceptions, the differences in perception between individuals, and the resulting difficulty of trying to integrate these perceptions. Despite these challenges, the perceptual domain offers the great advantage of being able to alter perceptions through communications, easily search for creative alternatives if needed, and make changes with minimum efforts and costs. The perceptual domain, however, relies on the virtual domain, and in some cases on the physical domain, for communication and validation.

7.9 References

7-1. Chrysler Corporation, Ford Motor Company, and General Motors Corporation. *Advanced Product Quality Planning (APQP) and Control Plan Reference Manual*, Issued June 1994.

7-2. Khurana A., and Rosenthal, S. "Integrating the Fuzzy Front End of New Product Development," *Sloan Management Review*, winter, 1997.

7-3. Khurana A., and Rosenthal, S., "Toward holistic 'front ends' in new product development," *J. Product Innovation Management* 15(1):57–74, 1998.

7-4. https://www.shutterstock.com/, Images 235877185 and 331112708.

7-5. https://www.shutterstock.com/, Image 331112780.

7-6. Cooper, R., and Dreher, A. "Voice-of-Customer Methods: What is the Best Source of New-Products Ideas?" *Marketing Management*. 19(4):38–43, 2010.

7-7. http://asq.org/learn-about-quality/qfd-quality-function-deployment/overview/overview.html.

7-8. http://www.jdpower.com/.

7-9. Mann, R., Abbas, A., Kohl, H., and Görmer, M. 2010. Global Survey on Business Improvement and Benchmarking. Global Benchmarking Network: Berlin. Available from: http://www.globalbenchmarking.org/fileadmin/user_upload/GBN/PDF/2010_gbn_survey_business_improvement_and_benchmarking_web.pdf.

7-10. https://www.shutterstock.com/, Image 235877185.

7-11. https://www.shutterstock.com/, Image 219303295.

7-12. M. El-Sayed, T. Stawiarski, and J. El-Sayed "Multi-Attribute Balancing Process for Automotive Components," ASME Paper No. IMECE2011-65009, *Proceedings of the ASME 2011 International Mechanical Engineering Congress*, Denver, Colorado, 2011.

7-13. M. El-Sayed and J. El-Sayed "Resolving Conflicting Product Attributes," Paper#: 30134, *ICMAAE 2013 Conference*, July 2013.

7-14. https://www.shutterstock.com/, Image 376256434.

7-15. https://www.shutterstock.com/, Image 219313252.

7-16. http://image.shutterstock.com/display_pic_with_logo/726964/726964,1329839266,1/stock-photo-transparent-car-concept-made-in-d-graphics-95719666.jpg.

7-17. http://articles.sae.org/4360/.

7-18. http://www.sae.org/dlymagazineimages/web/516/4360_4015.jpg. (Reprinted with permission from Ford.)

7-19. Kulkarni, A. "Development of an EV Drivetrain for a Small Car," Ph. D. Thesis, Faculty of Engineering and Industrial Sciences, Swinburne University of Technology, October 2013.

7-20. http://www.sae.org/dlymagazineimages/web/516/4360_4015.jpg. (Reprinted with permission from Ford.)

Chapter 8
Vehicle Virtual Realization

"Computers are to design as microwaves are to cooking."

—Milton Glaser

8.1 Introduction

Before the advancement in computational hardware and software, vehicle virtual realization was accomplished through engineering drawings and mockups. The advancements in computers and computational technologies did not only improve and facilitate virtual realization, but have been continuously expanding the virtual domain and rapidly changing realization processes, including the product realization process [8-1], [8-2], [8-3], [8-4].

During the product realization process, the perceptual realization phase is followed by the virtual realization phase. Like perceptual realization, virtual realization subphases are repeats of the three realization phases. Consequently, for virtual realization, the first subphase is virtual product and production process inception, the second subphase is virtual product and production process conception, and the third subphase is virtual product and production process maturation. In this chapter, these virtual realization phases are discussed. The main objective is to illustrate the steps for transforming a perceptual product concept to virtually developed and validated product and production process concept.

8.2 The Virtual Phase in New Product Realization

The virtual phase is the second phase in new product realization. As a realization phase, virtual realization follows the three subphases of inception, conception, and maturation, as shown in Figure 8.1.

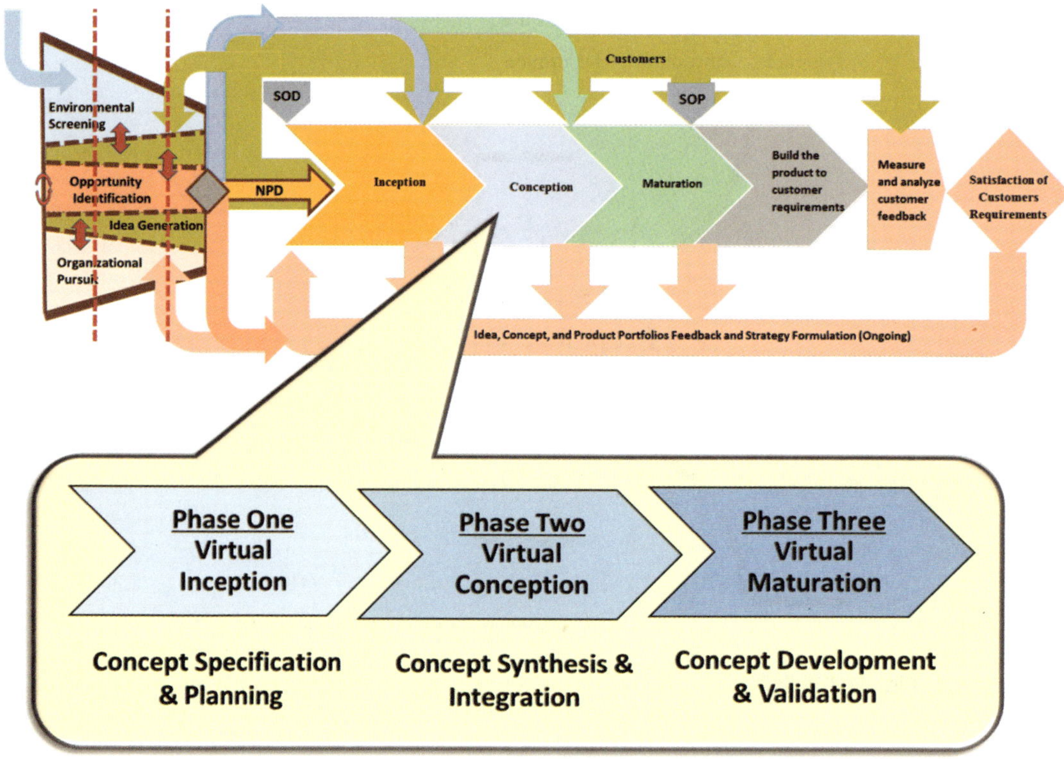

Figure 8.1 Virtual realization as the second phase of the new product realization process.

The first subphase is virtual inception or concept enrichment and specification for both product and production process. The second subphase is virtual conception or product and production process synthesis and integration. The third subphase is virtual maturation or product and production process development and validation.

The inception subphase is the beginning of virtual realization and the interface with perceptual realization. Therefore, it could be simultaneously performed and integrated with the validation activities of perceptual maturation. This integration should be the normal practice to accelerate the program unless it is halted during concept initiation approval because of some unforeseen issues. With the integration of the validation activities of perceptual maturation and virtual perception subphase, the structured product realization process inception and conception can be illustrated, as shown in Figure 8.2.

During virtual realization, product concept is enriched, specified, integrated, developed, and validated as close as possible to the established targets with no or minimum negative gaps. Therefore, closing the negative and increasing the positive gaps is a key quality pursuit. For closing or minimizing negative gaps during virtual realization, the product quality planning cycle shown in Figure 8.3 could be followed [8-5].

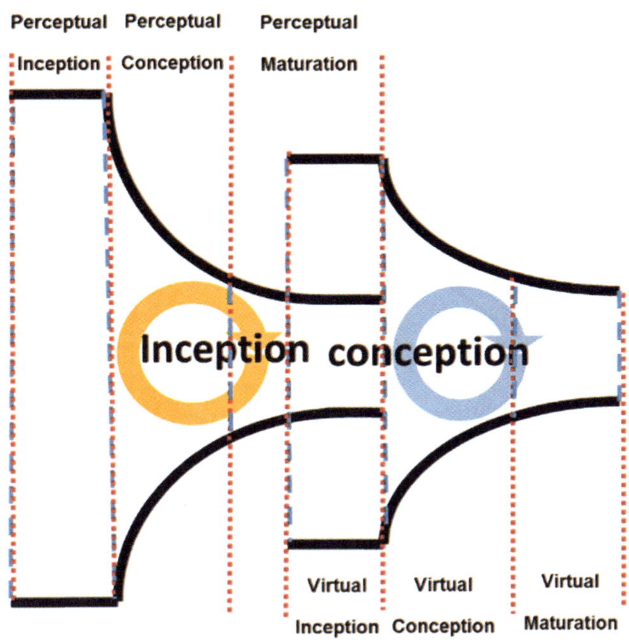

Figure 8.2 Joined perceptual validation and virtual inception activities in the product realization process.

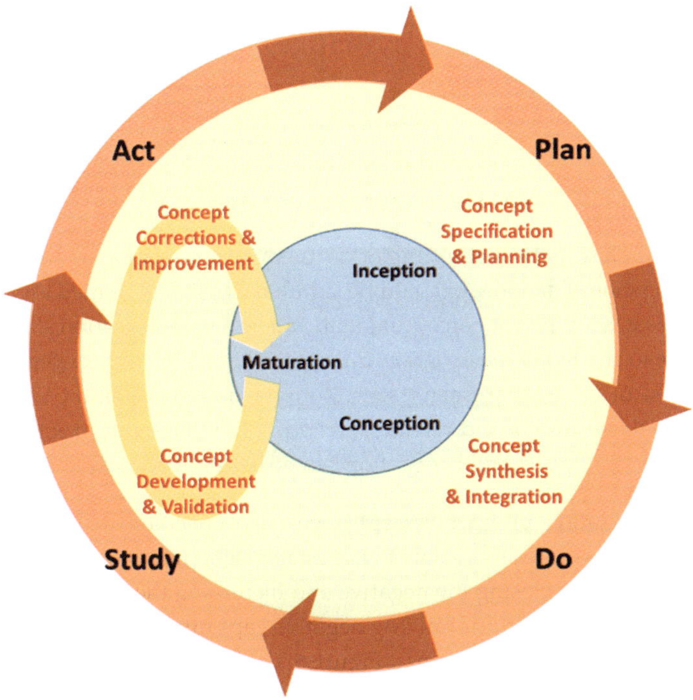

Figure 8.3 Virtual realization quality planning cycle.

Virtual realization gaps are mainly development gaps, contained in the array G_d, causing mostly linear potential losses to the producer. Therefore, during the virtual maturation iterations between virtual development/validation and corrections/improvements, all negative gaps should be addressed. These iterations should be continued until all necessary corrections and improvements are implemented and validated for both product and production process.

Product and production process should be virtually realized simultaneously or jointly from virtual inception to virtual maturation as two sides of the same concept. Due to product and process parameters' dependence, the best practice to optimize both product and process with realistic objectives, constraints, and priorities is joint development, as shown in Figure 8.4.

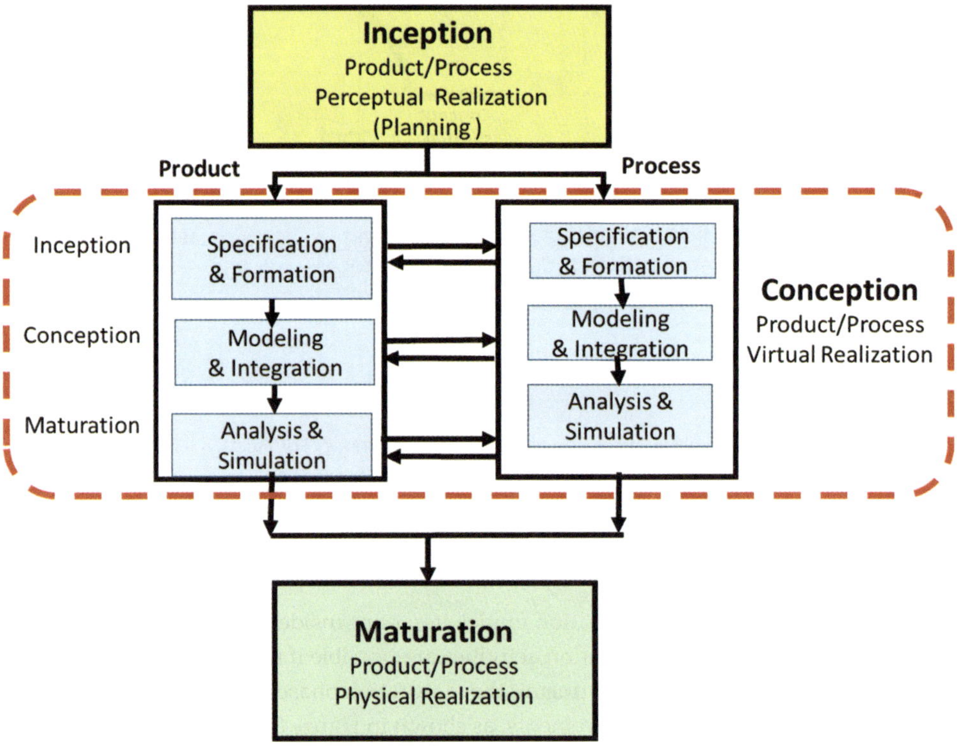

Figure 8.4 Joint virtual realization of product and production process.

8.3 Virtual Realization Model

Similar to the perceptual realization model, the virtual realization model shown in Figure 8.5 connects the perceptual realization outputs with the virtual inception, conception, and maturation. By conducting virtual enrichment and specification simultaneously with the activities of perceptual validation, virtual inception becomes part of phase one. Starting with phase two, virtual conception transforms the specified perceptual concept into a virtual product concept through synthesis and integration. Finally, the virtual realization is concluded by the virtual maturation of the product and production process concepts through development/validation and corrections/improvements iterations. The developed product concept is validated through virtual simulations, prototype testing, customer reviews, and business reviews. At the end of virtual realization, the decision to proceed to physical realization or send the virtual product and process concepts to the concept portfolio could be made by management and the cross-functional quality planning team.

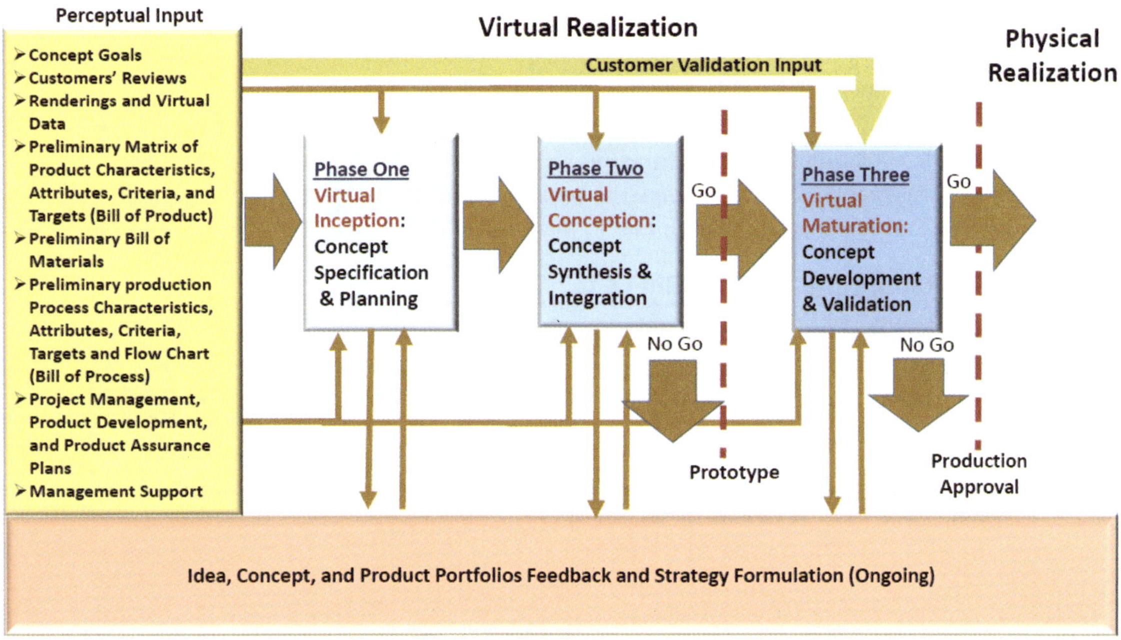

Figure 8.5 Virtual realization model.

As illustrated by the virtual realization model, iterations inside and between virtual inception, conception, and maturation activities are possible if needed. However, iterations are more likely to be needed during the maturation phase between development/validation and corrections/improvements, as shown in Figure 8.3, until all virtual realization issues are addressed. For instance, during virtual conception, any needed changes could be made without going back to virtual inception. During maturation, any validation issues can be addressed through further development without having to go back to conception. Customer and business reviews are conducted during the development and validation iterations until a decision is made to proceed to physical realization or the virtual concept is sent to concept portfolio. In the following, the key inputs and outputs in addition to the inception, conception, and maturation activities of the virtual realization model, shown in Figure 8.5, will be discussed.

8.4 Virtual Realization Inputs

Virtual realization starts with the following inputs from perceptual realization:

8.4.1 Concept Goals

The concept goals are the cumulative output of the realization activities from preinception to perceptual realization. These goals specify the desired outcomes of the new product realization endeavor. Communicating these goals during the transition can provide a guide for achieving and assessing virtual realization outcomes.

8.4.2 Customers' Reviews

During the iteration of perceptual maturation, quality planning cycle customers' reviews, assessment, and comments must be captured. Regardless of what assessment tools are utilized, the customers' reviews must be compiled in a format that is conducive to the implementation of meaningful improvements. For example, a useful format could highlight the strengths of the concept, the necessary corrections, and improvements, and any useful insights that could be shared.

8.4.3 Renderings and Virtual Data

All styling records, renderings, drawings, virtual simulations, and visualization data of the perceptually realized concept must be part of the inputs to the virtual realization domain. These data should contain the accumulated records of the preinception and perceptual realization phases.

8.4.4 Preliminary Matrix of Product Characteristics, Attributes, Criteria, and Targets (Bill of Product)

Along with perceptual concept goals, the matrix of product characteristics, attributes, criteria, and targets are the producer's perception of the voice of the customer that guides and assesses the realization process outcomes. The name bill of product was coined for this matrix because the matrix describes what the product should be in detail. In addition, this matrix should be the source of all technical specification from vehicle level to component levels, as will be discussed later.

8.4.5 Preliminary Bill of Materials

The preliminary bill of materials contains the list of the components and subsystems for the perceptually realized concept. This list should identify the off-the-shelf and the new subsystems and components to be designed and developed. Accordingly, the bill of materials must be part of the inputs to the virtual realization phase.

8.4.6 Preliminary Production Process Characteristics, Attributes, Criteria, and Targets (Bill of Process)

The production process preliminary matrix, or the bill of process, represents the perceptually realized characteristics, attributes, criteria, and targets needed to achieve the product based on the voice of the customer. The bill of process serves as the source for process form, function, and fit technical specifications. In addition to the preliminary bill of process, the perceptually realized and approved preliminary process flowchart must be communicated to enable process virtual realization.

8.4.7 Project Management, Product Development, and Product Assurance Plans

By identifying the tasks needed for virtual and physical realization of the product and production process, the product development plan can be completed during the maturation phase of perceptual realization. In addition, the product assurance plan can be concluded by identifying the fit characteristics, attributes, criteria, and targets. Once these plans are solidified, the project management plan can be developed as well. As key perceptual realization outputs, these plans and any insights must be communicated to all subsequent phases of product and process realization.

8.4.8 Management Support

Management support must be secured during the program approval phase before the start of virtual realization. To secure management support, project feasibility must be decided by the cross-functional team after reviewing the perceptually realized product and production process concept.

8.5 Virtual Inception

As the beginning of virtual realization, virtual inception is the interface with perceptual realization. As shown in Figure 8.6. virtual inception is simultaneously performed and integrated with the validation activities of perceptual maturation unless concept development was suspended. Guided by feedback from customers' reviews, perceptual realization validation, and portfolios' input, the product/process perceptual concepts can be enriched and all necessary changes and improvements can be made. As the perceptual maturation activities are tapered down, the virtual inception activities are ramped-up, as shown in Figure 8.6.

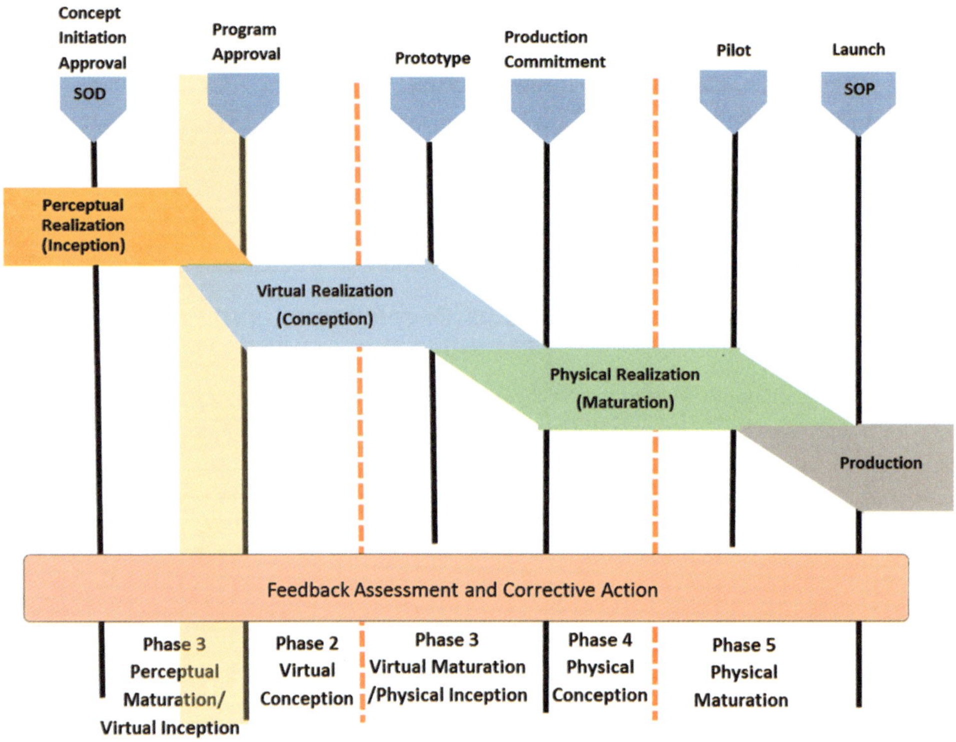

Figure 8.6 Virtual inception phase of new product realization process.

8.5.1 Product

During virtual inception, final improvements are introduced to the bill of product in the form of the full matrix of specific product characteristics, attributes, criteria, and targets, as discussed in Chapter 4. Once the bill of product is finalized, the vehicle technical specification (VTS) can be developed. Using the VTS and the preliminary bill of materials, the preliminary subsystem technical specifications (SSTS) and, subsequently, the preliminary component technical specification (CTS) are created. Building on the perceptual realization house of quality discussed in Chapter 6, the derivation of the technical specifications of the vehicle, subsystems, and components can be facilitated, as shown in Figure 8.7.

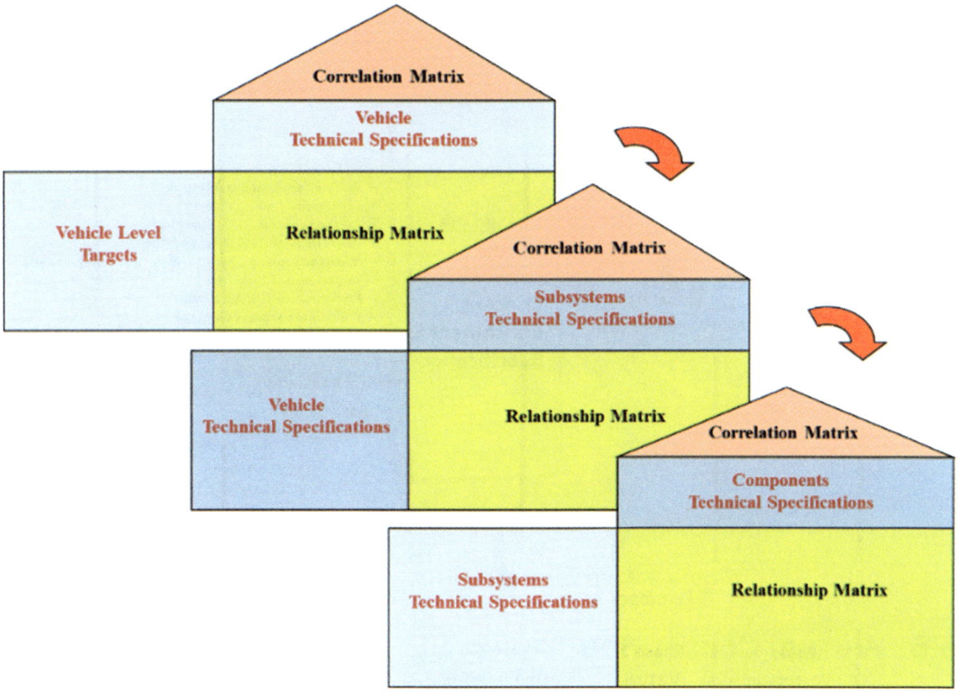

Figure 8.7 The house-of-quality sequence from vehicle-level targets to CTS.

8.5.2 Process

Similar to product specification, the initial full matrix of process characteristics, attributes, criteria, and targets can be derived, as discussed in Chapter 3. Once the bill of process is defined, the process flow and layout can be specified.

With product and production process targets and specification defined, the modeling and simulation activities for virtual realization can be planned. Consequently, the output list of virtual inception, as shown in Figure 8.8, include the following:

- vehicle technical specification
- subsystems technical specification
- components technical specification
- renderings, virtual data, and changes
- matrix of product characteristics, attributes, criteria, and targets (bill of product)
- preliminary bill of materials
- preliminary process characteristics, attributes, criteria, targets, specifications (bill of process), and flow chart
- project management, product development, and product assurance plans

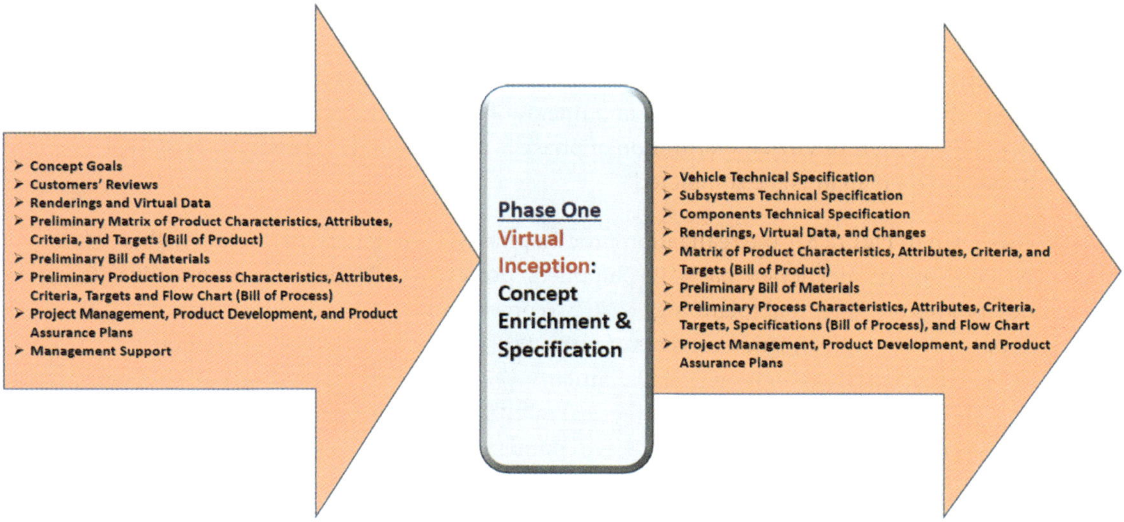

Figure 8.8 Virtual inception inputs and outputs.

8.6 Virtual Conception

With the integration of virtual inception and perceptual validation, as shown in Figure 8.9, virtual conception becomes phase two of the new product realization process.

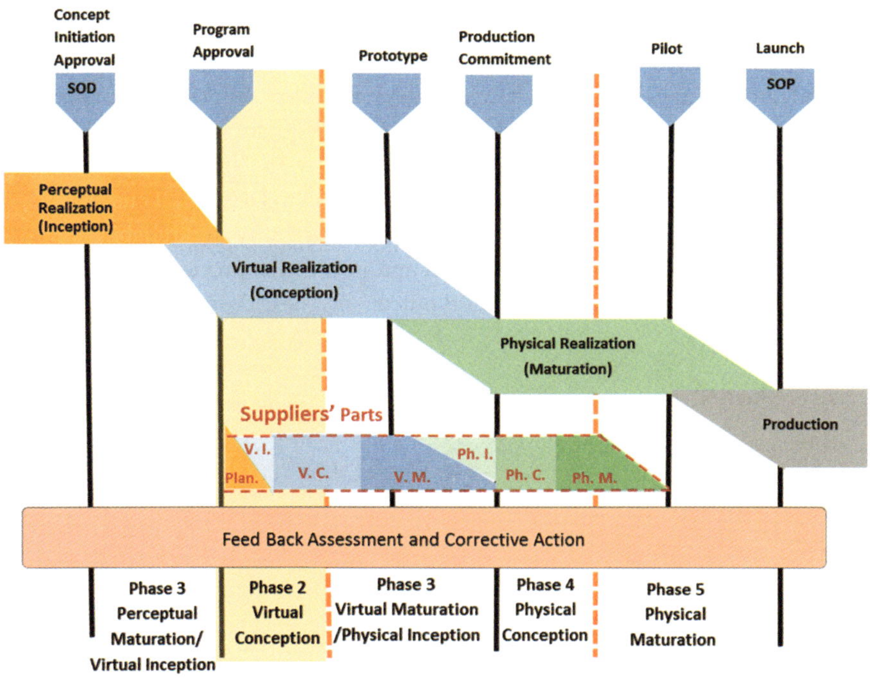

Figure 8.9 Virtual conception phase of the new product realization process.

Program approval marks the start of virtual conception. The specified product/process perceptual concept is transformed into a virtual product/process concept during virtual conception. Therefore, virtual conception marks the beginning of the product and production process virtual synthesis and integration. Before the end of virtual conception and the start of virtual maturation of phase 3, indicated in Figure 8.9, the prototype build control plan must be completed.

As shown in Figure 8.9, the realization process phases for suppliers' parts start with program approval. For new parts, the suppliers' process could follow the same realization phases of perceptual (planning), virtual (conception), and physical (maturation). Suppliers' virtual realization consists of virtual inception (V. I.), virtual conception (V. C.), and virtual maturation (V. M.). Similarly, physical realization consists of physical inception, physical conception, and physical maturation. For an integrated supplier, as shown in Figure 8.9, planning and V. I. are conducted simultaneously. When planning is completed, the V. I. of the supplied part moves to V. C.

While virtual conception activities are mainly design activities, for a mass-produced product virtual realization is complete only when both the product and process are fully developed and validated virtually. Therefore, product and process designs do not constitute full virtual realization until validated. Consequently, all unresolved issues and remaining conflicting targets from perceptual realization must be addressed during virtual conception. These issues and conflicts can be addressed using synthesis and integration tools such as multiobjective optimization or iterative virtual simulations [8-6].

During product virtual conception, the design for X (DFX) activities to achieve the desired specified product form, specified targets, attributes, and characteristics begin with architectural integration and virtual modeling. For a mass produced product, design for manufacturability (DFM) [8-7], [8-8], [8-9], design for assembly (DFA) [8-10], [8-11], design for six sigma [8-12], [8-13], design sensitivity analysis [8-14], [8-15], [8-16], and robust design [8-17], [8-18] are conducted in addition to DFX for all customers' desired characteristics and attributes, such as safety, durability, and reliability. Performing the virtual conception of product and production process simultaneously is the key to achieving DFM and DFA desired outcomes. These desired outcomes include the following:

1. cost reduction
2. minimum part variations
3. ease of fabrication
4. minimum number of parts
5. minimized complexity
6. modular subassemblies

7. ease of assembly
8. minimized complexity
9. minimized handling
10. optimized flow and layout

To demonstrate the DFX activities and outcomes for form and function, including manufacturability and assembly, during virtual conception, the following case study is considered.

8.6.1 Case Study 8.1: Virtual Conception of a Door System for Styling, Safety, and Manufacturability

To achieve five-star side impact rating, while meeting styling, manufacturability, and other functionality requirements, perceptual and virtual realization in addition to physical concept proofing of a modular door system were conducted. During inception—besides specifying styling, manufacturability, assembly, and other functional requirements—the design criteria, targets, and specifications for five-star rating were established according to FMVSS-214 [8-19]. The rating includes thoracic trauma index, pelvic acceleration, and maximum door deformation. By decoupling form and functional requirements, a modular space frame perceptual concept was developed. During the virtual conception synthesis, integration, DFX, DFM, and DFA were performed. In addition to crafting self-contained modules that integrate door latch, window regulator, and hinges, different materials and formability options, such as hydroforming (Figure 8.10) and stamping (Figure 8.11) were studied [8-20], [8-21], [8-22].

Figure 8.10 Hydroformed door module concept.

Figure 8.11 Stamped door module concept.

During virtual conception, synthesis and integration activities may include some physical builds and simulations such as mockups, test samples, and data acquisition vehicles, subsystems, and/or parts. Examples of these physical builds include interior and exterior mockups for styling, partial builds for crash safety testing, and load acquisition mule for durability. The physical simulation builds such as mockups and test samples can be considered as exploratory localized mapping to the physical domain. The data acquisition physical builds can be considered as reverse localized mappings from the physical domain to guide the realization process in the virtual domain. In the following, some selected product/process form, function, and fit activities will be discussed to illustrate the virtual conception flow.

8.6.2 Product Form

Product virtual conception activities are mainly design activities that shape the final outcome of both the product and production process. For exterior and interior styling, computer-aided styling (CAS), as exemplified by Figure 8.12 [8-23], and clay modeling are utilized. Also, architectural integration modeling using computer-aided design (CAD) tools in addition to tolerance studies with geometric dimensioning are performed.

Figure 8.12 Wireframe of a sports car concept [8-23].
(Mikhail Bakunovich/shutterstock.com)

8.6.3 Product Function

To lead and guide product virtual conception, early function analysis using simplified virtual models and simulation tools can be utilized. These virtual simulations could help in performing tradeoff studies and conduct functional, cost, and risk analyses. Figure 8.13 [8-24] provides an example of a suspension 3-D multibody modeling and simulation during virtual conception.

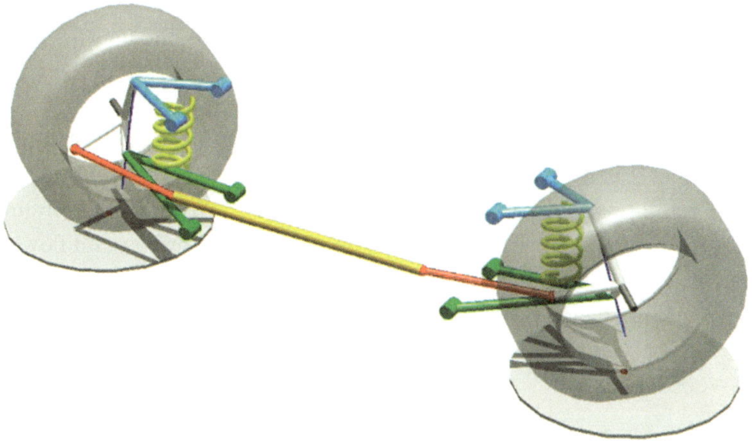

Figure 8.13 Suspension 3-D multibody modeling and simulation [8-24].

8.6.4 Process Form

Process architectural integration and modeling using digital manufacturing tools are performed during virtual conception to meet established process targets. The main drive for process virtual conception activities is to optimize the production process layout and flow by aligning operations and stations. The virtual facility layout and flow, as illustrated by the 2-D virtual conception in Figure 8.14 [8-25], is usually optimized based on the sequence of activities, the flow of materials, and available real estate.

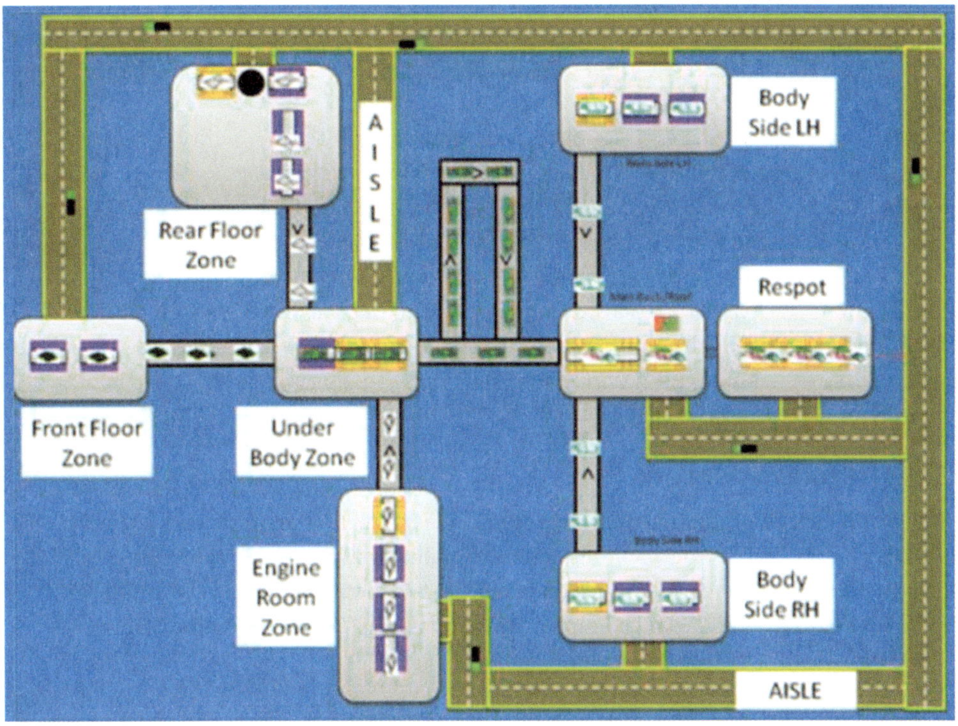

Figure 8.14 2-D virtual conception of a plant layout and flow [8-25].

During virtual conception, process DFX activities to achieve desired specified process form, targets, attributes, and characteristics begin with architectural integration and virtual modeling. The process flow and initial plant layout start with the available real estate, product and process virtual data, preliminary bill of materials, preliminary production process characteristics, attributes, targets, specifications, and flowchart. The process specifications and integration are established based on product/process assumptions, benchmarking, historical warranty data, quality/dependability data, reliability studies, test requirements, operation sheets, annual production rate, the number of shifts, and work schedule.

8.6.5 Process Function

Process functional conception shapes the final outcome of both the product and production process characteristics and attributes. Process functional integration, DFX activities, modeling, and simulation using CAD/CAM/CAE, digital manufacturing, and 3-D virtual visualization tools should lead and guide process virtual conception to ultimately meet the established process targets and specifications.

During virtual conception, virtual simulations and visualization could facilitate selection of manufacturing strategies, evaluation of alternatives, reduction of variance, coordination of production resources, and estimation of throughput and costs.

8.6.6 Product/Process Fit

Product and process failure mode and effect analysis (FMEA) are quality tools for mitigating possible product and process failures, uncertainty, and risks. To account for unconsidered criteria, an early virtual FMEA, usually called design FMEA (DFMEA), is performed. Accounting for potential failures, using DFMEA during virtual conception assures product quality and minimizes the number of development and validation iterations during virtual maturation. To demonstrate the use of simplified virtual simulation tools in detecting potential product/process failures, the following case study is considered.

8.6.7 Case Study 8.6.3: Thinning of an A-Pillar Section Due to Formability [8-3]

Thinning due to formability could have many adverse effects, such as localized fatigue failure, because of the crack initiation life degradation. However, in some cases, thinning may not be considered as part of the design criteria. To detect thinning during conception, simple plane strain formability simulations could be used. Figure 8.15 shows the locations of different body sections, including an A-pillar section where thinning simulations could be performed. Figure 8.16 shows a simplified plane strain finite element model used in the formability simulation of an A-pillar section, while Figure 8.17 shows the thinning percentage contours for the simulated section [8-3]. Using a block of a typical load history for the A-pillar section, the fatigue life reduction can be estimated at the worst thinning locations [8-26].

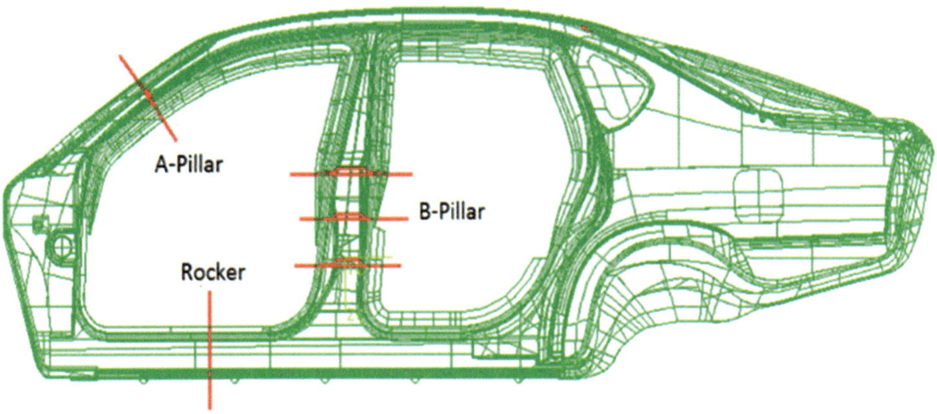

Figure 8.15 Locations of different side panel sections for formability simulation [8-3].

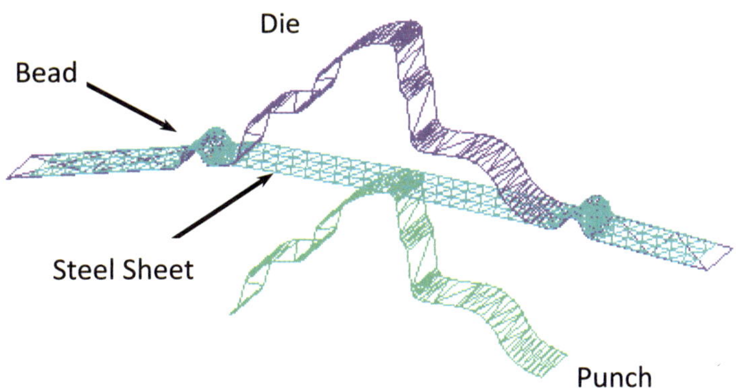

Figure 8.16 The plane strain formability model for an A-pillar section [8-3].

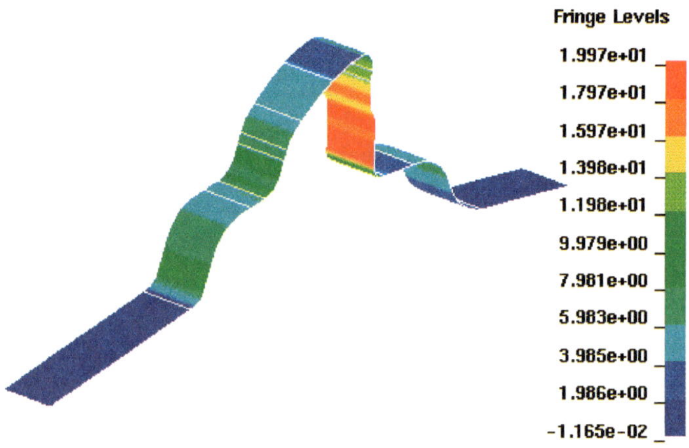

Figure 8.17 Percentage thinning contours for simulated A-pillar section [8-3].

To monitor product/process integration progress during virtual conception, periodic, planned, and documented design reviews must be conducted by the cross-functional team. These reviews may encompass product/process DFX activities and results, DFMEA, virtual modeling and simulation results, tradeoffs, technical and timing status, and progress relative to time constraints. Finally, prototype build control and test plans listing geometric, material, and functional tests must be completed by the end of virtual conception phase. Accordingly, the output list of virtual conception, as shown in Figure 8.18, include the following:

- virtual product/process integration reviews
- product/process virtual modeling and simulations (DFX, FMEA, DFM, DFA …)
- product/process math data and engineering drawings
- vehicle, subsystems, and components technical specification (technical specifications)
- material specifications
- matrix of product characteristics, attributes, criteria, and targets (bill of product)
- bill of materials
- preliminary process characteristics, attributes, criteria, targets, specifications (bill of process), and flow chart
- prototype build, control, and test plans
- project management, product development, and product assurance plans

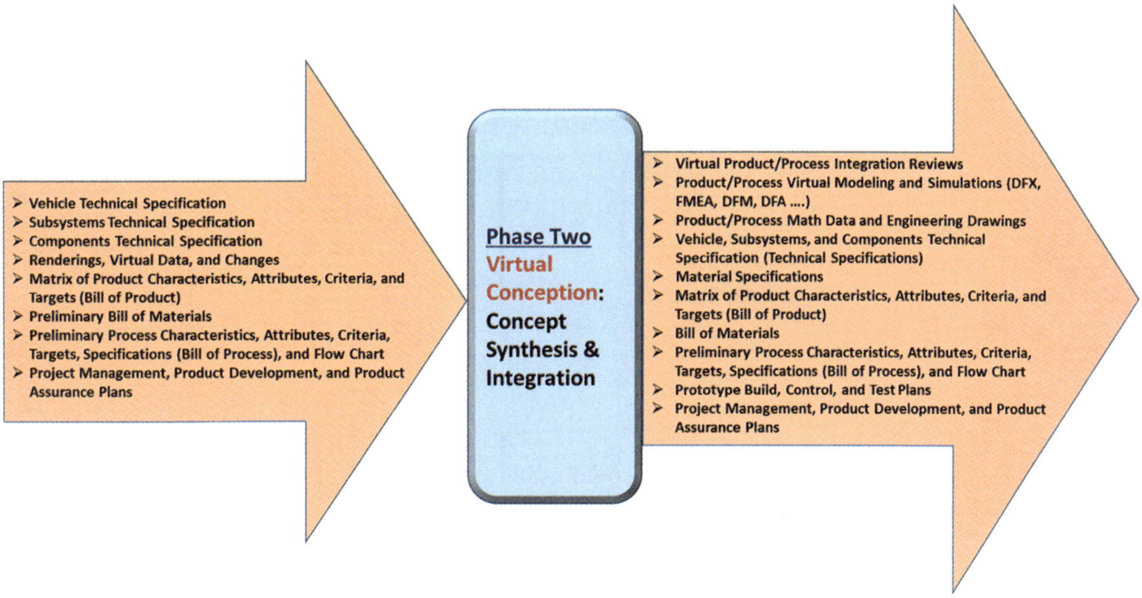

Figure 8.18 Virtual conception inputs and outputs.

8.7 Virtual Maturation

Product/process virtual development and validation are the main activities during virtual maturation. While virtual conception is the start of the product/process virtual synthesis and integration, virtual maturation or phase 3, as shown in Figure 8.19, is the virtual development/validation and corrections/improvements phase. During virtual maturation, iterations are conducted to develop, correct, improve, and validate the product/process virtual conception. For supplied parts, as shown in Figure 8.19, suppliers' virtual maturations may begin after the start of product/process virtual maturation but should be concluded during the same time frame.

One of the key benefits of identifying the three realization phases of perceptual, virtual, and physical is understanding the nature of each phase and the mapping between them. As discussed earlier, modeling and simulations as basic elements for validation can be perceptual, virtual, or physical. Therefore, validation during each phase could be performed using the same type of modeling and simulations for that phase or the next phase, if needed. For example, virtual validation can be performed using virtual modeling and simulation tools if these tools are available; otherwise, physical modeling and simulation may be needed.

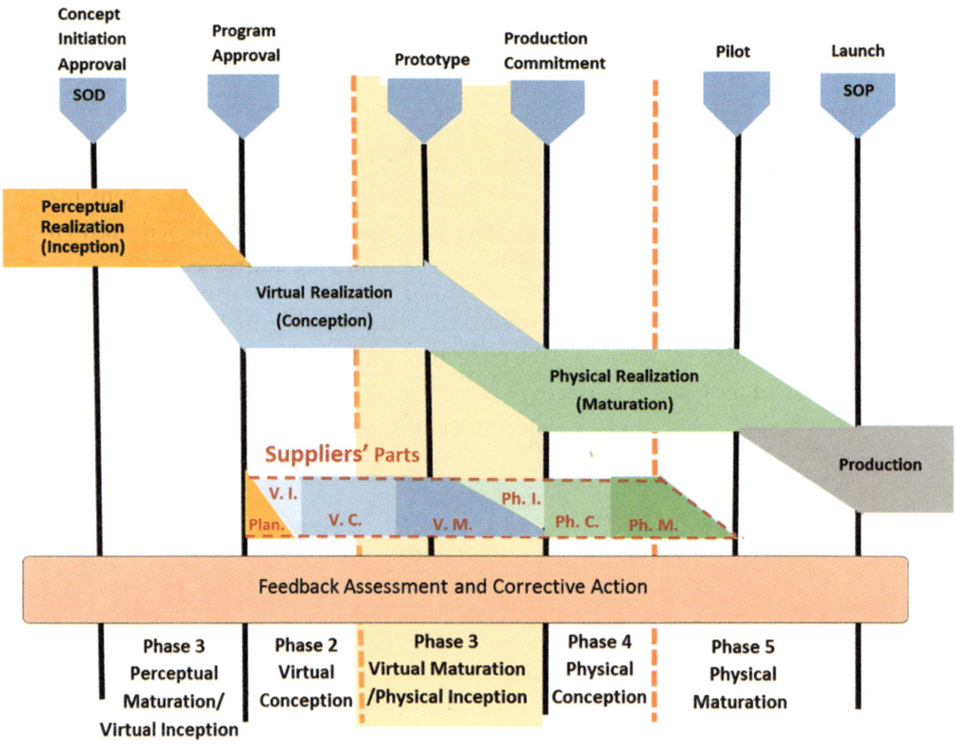

Figure 8.19 Virtual maturation phase of the new product realization process.

For form validation, using physical builds during virtual maturation provides a preview of the mapped virtually realized concept into the physical domain. For functional validation, using physical modeling and simulations during virtual validation provides a confirmation that the virtually realized performance could be achieved in the physical domain. Physical builds during virtual realization, however, are mostly handcrafted and assembled in small numbers using unique tools. Consequently, these builds are very costly and time-consuming. Furthermore, during virtual maturation, physical simulations do not validate the product or process but only validate the virtually realized concept or design. Therefore, for functional validation during virtual maturation, physical builds should be limited to functions with unavailable virtual simulation tools or lack established correlations between the physical and virtual domains. Accordingly, to reduce product realization process time and cost, continuous improvement of virtual simulations tools and development of reliable correlations between physical and virtual performances should be supported for the following advantages [8-3]:

- ability to examine several alternatives and provide direction to optimize product/ process
- better understanding of the design space and its constraints
- fast turnaround time and ability to lead the realization process

- cost effectiveness and high return on investment
- shorter product realization cycle

8.7.1 Product Form

The virtual product form maturation begins with development activities to achieve desired specified product form targets, attributes, and characteristics. Development/validation and corrections/improvements iterations are initiated and continued until virtual realization is completed, and all virtual maturation issues are addressed.

CAS and clay modeling activities are continued as shown in Figure 8.20. These styling activities are concluded by achieving the curvature continuity and tangency alignment targets established for Class A surfaces [8-27]. By achieving the established targets, the desired Class A surfaces aesthetical reflection quality is accomplished.

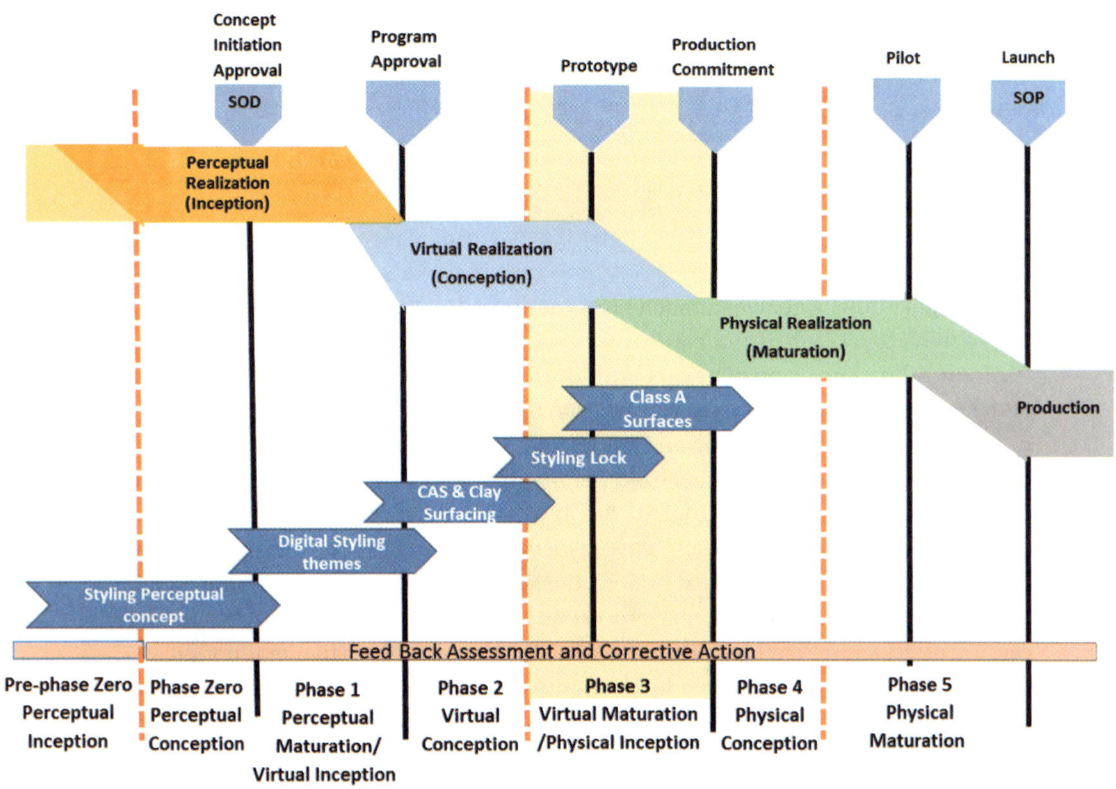

Figure 8.20 Styling progress during realization phases.

During product virtual maturation, planned design verification activities are conducted to confirm that tolerances are according to industry standards and can be manufactured. As the design is being matured and virtual development, corrections, improvements, and validation iterations are diminished, engineering drawings, math data, and material specifications could be finalized and ultimately approved.

Form virtual development for components, subsystems, and the full vehicle must be completed by prototype build. In addition, all interface issues must be resolved. For form virtual validation, few proof-of-concept physical prototypes could be built for visual evaluations and customers' viewing.

8.7.2 Product Function

Virtual performance simulations of parts and subsystems in the full-vehicle context are performed during early development activities of virtual maturation. Development activities are finalized with the achievement of targeted performances. For key components and functions, virtual performance simulation targets should be based on pre-established reliability and confidence levels with strong correlation to customers' physical realities. Depending on the availability of virtual modeling and simulation tools, most functional validation during virtual maturation can be accomplished without physical builds. However, as discussed earlier, physical builds during virtual maturation could be used for functions with unavailable simulation tools or unestablished correlations between the virtual and the physical domains.

Full-vehicle virtual simulations could be used during iterations between corrections/improvements and development/validation until virtual realization is completed and all performance issues are addressed. Examples of such simulations are illustrated by the nonlinear finite element simulations of a pothole and off-road full-vehicle durability, shown in Figure 8.21 [8-3] and Figure 8.22 [8-3], respectively.

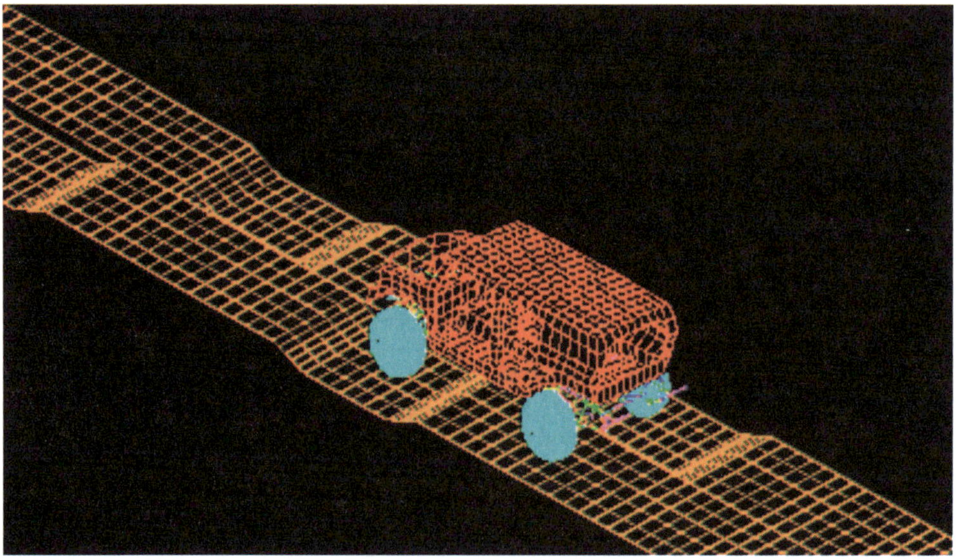

Figure 8.21 Virtual full-vehicle pothole durability simulation [8-3].

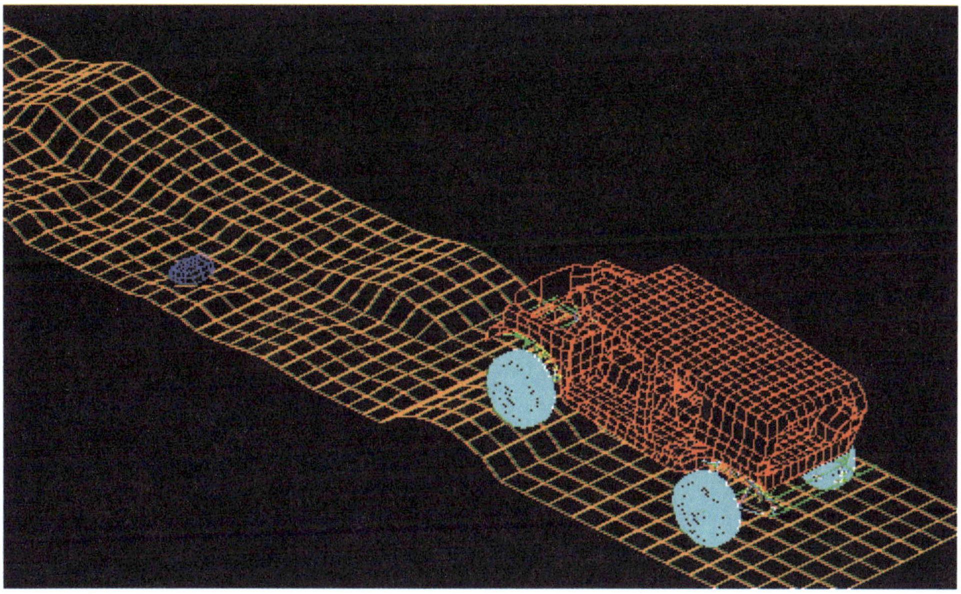

Figure 8.22 Virtual full-vehicle off-road durability simulation [8-3].

8.7.3 Process Form

The virtual process form maturation begins with development activities to achieve the desired specified process form targets, attributes, and characteristics. Subsequently, development/validation and corrections/improvements iterations are initiated and continued until the process virtual maturation is completed and all validation issues are addressed. In addition to virtual simulations tools, the experiences and lessons learned from previous physical builds could aid during process form development, verification, and validation.

During process virtual maturation, process design verification activities should be conducted to confirm continuity and meeting all safety and operational standards. Using 3-D modeling and simulation, as exemplified by Figure 8.23 [8-28], a realistic process form can be achieved. As the process design is matured through the virtual development and validation iterations, the requirements for new facilities, equipment, tools, and gauges could be accurately defined, finalized, and ultimately approved.

Figure 8.23 3-D model of an overhead vehicle assembly system [8-28]. (Macrovector/shutterstock.com)

8.7.4 Process Function

The virtual process functional maturation begins with development activities to achieve the desired process performance targets, attributes, and characteristics. During development activities, performances of different process operations such as ergonomics, robotics, welding, painting, and assembly could be modeled and simulated virtually.

Quality cycle development/validation and correction/improvement iterations for both product and production process should be continued until all performance targets are achieved and an integrated product/process virtual realization is validated. As illustrated in Figure 8.24, assembly simulation and validation can be performed using virtual modeling, simulation, and validation tools [8-29].

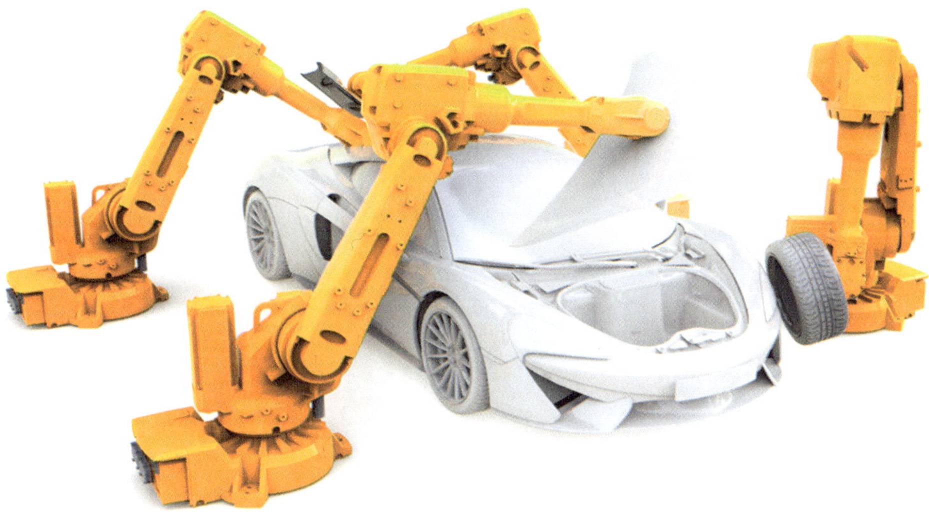

Figure 8.24 Product and production process virtual assembly simulation and validation [8-29]. (Solcan Design/shutterstock.com)

8.7.5 Product/Process Fit

In addition to finalizing virtual product/process FMEA and critical-to-quality analyses, design for six sigma and design of experiments should also be conducted and concluded before the end of virtual maturation. For successful mapping into the physical domain, key product and process characteristics should be clearly recognized. In addition, design space exploration and sensitivity analyses in the physical domain are costly, time-consuming, and provide limited insights. Therefore, clear identification of key components and subsystems' design spaces and sensitivities must be well explored and documented during virtual realization. In addition, with the flexibility and deeper insights provided in the virtual domain, gauge/testing requirements can be clearly identified.

To monitor product/process development and validation during virtual maturation, the periodic, planned, and documented design reviews must be conducted by the cross-functional team at least once during every iteration. These reviews should focus on the maturation activities, development progress, validation results, proposed corrections, and necessary improvements. Ultimately, the team must be convinced that the virtually realized product/process concept meets desired customers' characteristics, attributes, targets, specifications, and ready for physical realization and mass production. The virtual maturation iterations could be concluded as soon as the cross-functional team members are satisfied, production feasibility is certain, and the team feasibility statement sign off is accomplished. Consequently, the output list of virtual maturation, as shown in Figure 8.25, includes the following:

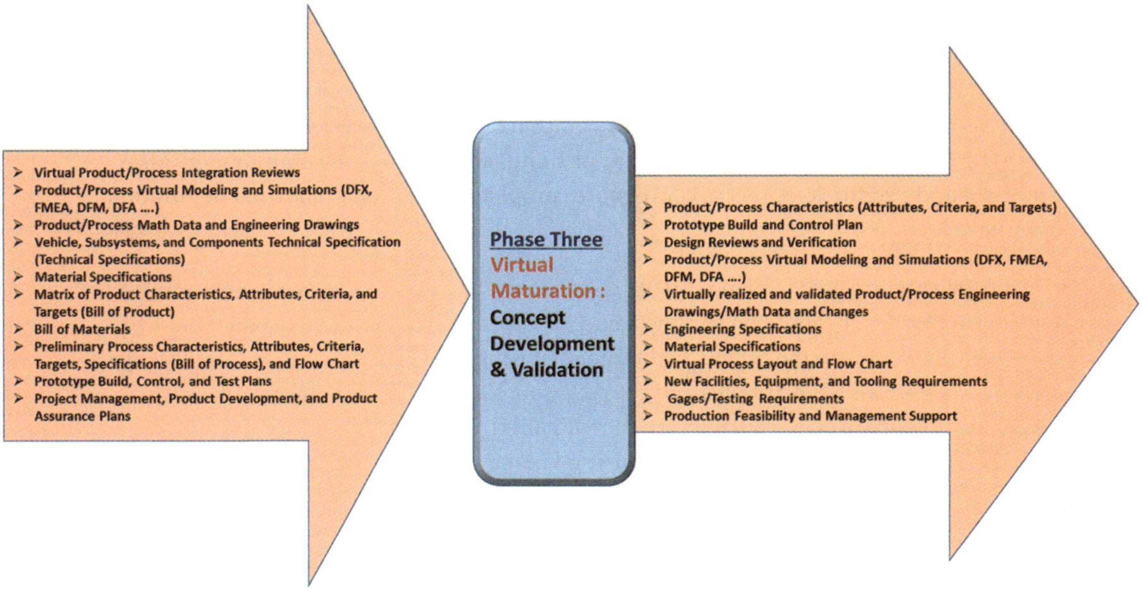

Figure 8.25 Virtual maturation inputs and outputs.

- product/process characteristics (attributes, criteria, and targets)
- prototype build and control plan
- design reviews and verification
- product/process virtual modeling and simulations (DFX, FMEA, DFM, DFA …)
- virtually realized and validated product/process engineering drawings/math data and changes
- engineering specifications
- material specifications
- virtual process layout and flowchart
- new facilities, equipment, and tooling requirements
- gauges/testing requirements
- production feasibility and management support

8.8 Learning Outcomes from Vehicle Virtual Realization

The desired learning outcomes from studying vehicle virtual realization can be categorized as follows.

8.8.1 Knowledge

Virtual realization is the product and production process synthesis and integration, or the traditionally known design phase. Similar to any realization phase, the three identifiable virtual realization subphases are virtual inception, virtual conception, and virtual maturation. During virtual inception or phase 1, the product and process perceptual concept is enriched and further specified. The transformation of product and production process perceptual realization into a product and production process virtual concept through synthesis and integration begins with virtual conception or phase 2. Development, validation, prototype build, and production commitment of the product and production process concept are completed by the end of virtual maturation or phase 3.

To achieve quality goals during virtual realization, the product and production process perceptual concept should be enriched, specified, integrated, developed, and validated as close as possible to the established targets. Closing all the negative gaps from targeted form, function, and fit attributes and characteristics and increasing the positive gaps should be the main strive during virtual realization. This can be accomplished by conducting product/process DFMEA during virtual conception and through the iteration development/validation and corrections/improvements of the quality cycle during virtual maturation.

During the product/process virtual conception, DFX, DFM, DFA, design for six sigma, design sensitivity analysis, and robust design are conducted to achieve the established targets, attributes, and characteristics such as safety, durability, and reliability. The unresolved issues and remaining conflicting targets from perceptual realization can be addressed using virtual synthesis and integration tools such as multiobjective optimization or through iterative virtual simulations.

In addition to validating virtual realization with physical prototype builds, some limited physical modeling and simulations can facilitate the identification and accelerate the resolution of subsystems and components interface issues. Examples of these physical builds include mockups for styling and packaging, partial builds for crash safety testing, and load acquisition mules for durability.

8.8.2 Skills

As the middle realization phase, the virtual realization is the connection between perceptual realization and physical realization. Accordingly, virtual realization plays a pivotal role in achieving and shaping the desired product. Therefore, in addition to analytical and interpersonal skills, a high level of synthesis and integration skills is needed during virtual realization. Without full vehicle synthesis and integration skills and full product and process realization perspective, the advantages and insights provided by virtual simulations may be lost.

8.8.3 Paradigm/Attitude

The virtual domain offers great advantages and insights for the realization of both product and production processes. However, during virtual realization, physical simulations do not validate the product or process. They will only validate the virtually realized concept or design. Therefore, the cost and time of associated physical builds should be minimized as possible. In addition to seeking better and more representative virtual simulation tools, the key to minimizing physical builds during virtual simulations is to continuously establish and improve correlations between the physical and virtual domains.

8.9 References

8-1. El-Sayed, M., "Rethinking the Automotive Design and Development Processes for Product Realization," SAE Technical Paper 2008-01-0861, 2008, doi:10.4271/2008-01-0861.

8-2. El-Sayed, M., "Lean Design for Integrated Product Realization," *SAE Int. J. Mater. Manuf.* 3(1):194-201, 2010, doi:10.4271/2010-01-0400.

8-3. El-Sayed, M., "Expanding Virtual Simulation in Product Realization," *SAE Int. J. Mater. Manuf.* 4(1):788-798, 2011, doi:10.4271/2011-01-0530.

8-4. MTS-CD Version 1.1. 1999. Accelerated Vehicle Development. MTS Systems Corporation: Minnesota.

8-5. Chrysler Corporation, Ford Motor Company, and General Motors Corporation, *Advanced Product Quality Planning (APQP) and Control Plan Reference Manual*, Issued June 1994.

8-6. Song, D., and El-Sayed, M. "Multi-objective optimization for automotive performance," *International Journal of Vehicle Design* 30 (4): 291–308, 2002.

8-7. Bralla, J. 1998. *Design for Manufacturability Handbook*, 2nd Edition, McGraw Hill: New York.

8-8. Jacquel and Salmon. "Design for Manufacturability: A Feature-Based Agent-Driven Approach," *Proceedings of the Institution of Mechanical Engineers.* 214(Part B):865–879, 2000.

8-9. El-Sayed, M. and El-Sayed, J., "Balancing Manufacturability and Performance Attributes in Lean Design," *SAE Int. J. Mater. Manf.* 5(1):174-182, 2012, doi:10.4271/2012-01-0512.

8-10. Boothroyd, G. 2005. *Assembly Automation and Product Design*, 2nd Edition. Taylor and Francis: Boca Raton, Florida.

8-11. Boothroyd, G., Dewhurst, P., and Knight, W. 2002. *Product Design for Manufacture and Assembly*, 2nd Edition. Marcel Dekker: New York.

8-12. El-Haik, Basem, and Raid Al-Aomar. 2006. *Simulation-Based Lean Six-sigma and Design for Six-Sigma.* John Wiley & Sons: New Jersey.

8-13. Sokovic, M., D. Pavletic, and K. Kern Pipan. "Quality improvement methodologies—PDCA cycle, RADAR matrix, DMAIC and DFSS." *Journal of Achievements in Materials and Manufacturing Engineering* 43(1):476–483, 2010.

8-14. El-Sayed, M. E. M., and K. W. Zumwalt. "Efficient Design Sensitivity Derivatives for Multi-Load Case Structures as an Integrated Part of Finite Element Analysis." *Computers and Structures* 40(6): 1461–1467, 1991.

8-15. Cameron, T., Thirunavukarasu, A., and El-Sayed, M. "Optimization of Frame Structures with Flexible Joints." *Structural and Multidisciplinary Optimization* 19(3):204–213, 2000.

8-16. Cameron, T., Jordan, L., and El-Sayed, M. "Sensitivity of Structural Joint Stiffnesses With Respect to Beam Properties: A Hybrid Approach." *Computers and Structures* 63(6):1037–1041, 1997.

8-17. Jen, E. 2005. *Robust Design*. Oxford University Press: Oxford.

8-18. Dehnad, K. 2012. *Quality Control, Robust Design, and the Taguchi Method*. Springer Science & Business Media: Berlin.

8-19. U.S. Department of Transportation. *National Highway Traffic Safety Administration Laboratory Test Procedure for FMVSS 214S (STATIC), Side Impact Protection*, Report TP-214S-05, April 15, 1992. Available from: http://www.nhtsa.gov/DOT/NHTSA /Vehicle Safety/Test Procedures/Associated Files/TP-214-s05.pdf.

8-20. Townsend, J., Kaczmar, M., and El-Sayed, M. "Modular Door System for Side Impact Safety of Motor Vehicles." International Technical Conference on Enhanced Safety of Vehicles, 2001, 2001-06-0167.

8-21. Townsend, J., and El-Sayed, M. "Modular Door System for Side Impact Safety of Motor Vehicles," Proceedings of the NHTSA conference, Amsterdam, June 2001.

8-22. El-Sayed, M., Townsend, J., and Kaczmar, M. "Automotive Space Frame Modular Door Systems for High Performance," Proceedings of HPSC Conference, Spain, March 2002.

8-23. https://www.shutterstock.com/image-illustration/red-sports-car-wireframe-nonbranded-concept-115233823?src=RMx94UR3lIWisBrXMWKUJA-1-13.

8-24. Alexander, S., Cozby, D., Foster, J., Henry, A., Houlihan, J., Kreppel, A., Pavlawk, R., and Wang, Y. "Redesign of a Road Course Racing Vehicle," Senior Vehicle Design Capstone (EL-Sayed, M., Instructor), Kettering University, December 2016.

8-25. Gangal, M., Thakur, A., Roy, D., Kumar, P. et al., "Integrated, Virtual Plant Design and Commissioning Methodology Using Digital Manufacturing and Lean Principles," *SAE Int. J. Mater. Manf.* 2(1):368-383, 2009, doi:10.4271/2009-01-0609.

8-26. J. El-Sayed. "Evaluation of Fatigue Life Degradation due to the Metal Forming Processes." *WIT Transactions* 76:157–166, 2004.

8-27. Autodesk Tutorials, "Understanding Class A Modeling," http://help.autodesk.com/view/ALIAS/2015/ENU/?guid=GUID-64611955-D2CC-44F2-98F0-D4F1FE931D8B.

8-28. https://www.shutterstock.com/image-vector/automotive-overhead-chain-monorail-conveyor-vehicles-490588654.

8-29. https://www.shutterstock.com/image-illustration/3d-render-image-representing-industrial-robots-469228925?src=6iuuC5eSftiTV-1kKtVEYQ-1-12.

Chapter 9
Vehicle Physical Realization

"Success is the progressive realization of a worthy goal or ideal."

—Earl Nightingale

9.1 Introduction

For mass produced vehicles, the physical realization is the culmination of the perceptual and virtual realizations. As the last realization phase, the main focus during physical realization is the preparation of the product and production process for a successful launch. Some of the main competitive dimensions in the automotive sector are quality, reliability, and low cost [9-1]. Accordingly, organizational profitability, competitiveness, and market share depend on the success of the realization process in achieving desired product and process form, function, and fit characteristics during the last phase of physical realization.

Like perceptual and virtual realizations, physical realization subphases are repeats of the three realization phases. Consequently, for physical realization, the first subphase is physical product and production process inception, the second subphase is physical product and production process conception, and the third subphase is physical product and production process maturation. In this chapter, these physical realization subphases are discussed. The main objective is to illustrate the steps for transforming virtually realized product concept ("concept vehicle") into validated mass production process and mass-produced product.

9.2 Physical Realization in New Product Development

Physical realization follows the three subphases of inception, conception, and maturation, shown in Figure 9.1.

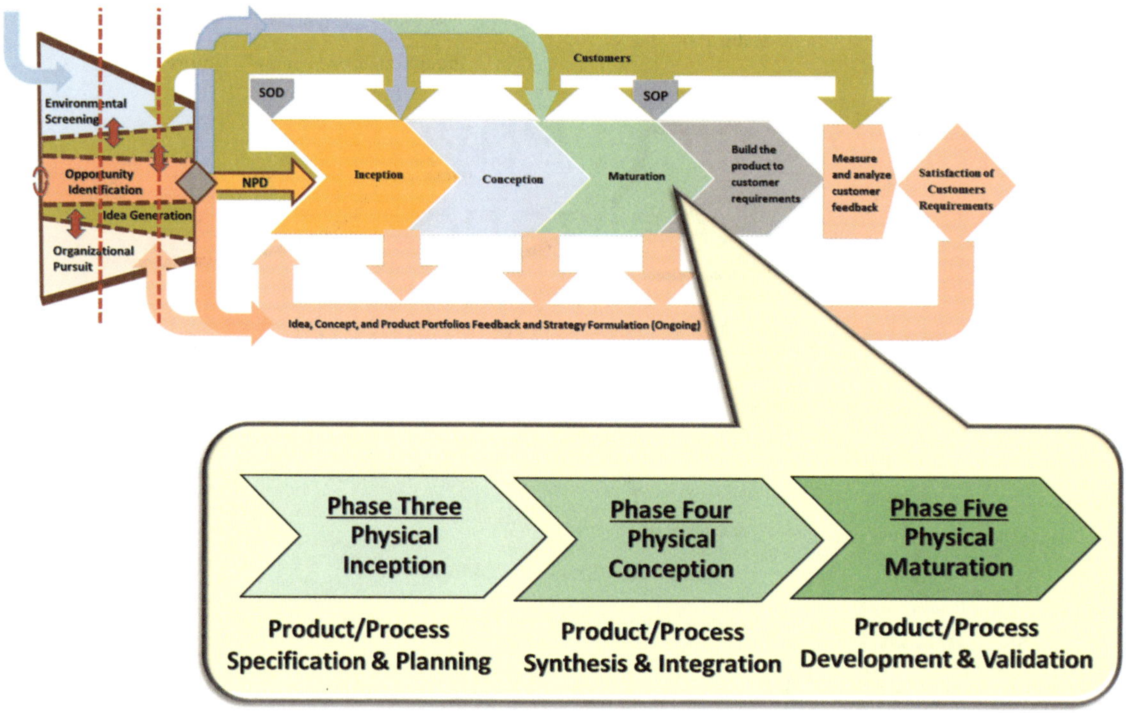

Figure 9.1 Physical realization is the third mapping of the new product realization process.

Physical realization is the mapping of a new product concept from the virtual domain to the physical domain. In new vehicle development, physical realization deals with product and process maturation from verified product and process virtual concepts to the fully realized and validated mass produced product and production system. The first subphase of physical inception deals with product/process physical specifications and planning. The second subphase of physical conception deals with product/process physical synthesis and integration. The third subphase of physical maturation deals with product/process physical development and validation.

The integration of physical inception with virtual validation (concept proofing and design verification) should be the normal practice to accelerate the program unless it is halted during prototype build approval because of some unforeseen issues. With the integration of the validation activities of virtual maturation and physical inception, the structured product realization process inception, conception, and maturation can be illustrated as shown in Figure 9.2.

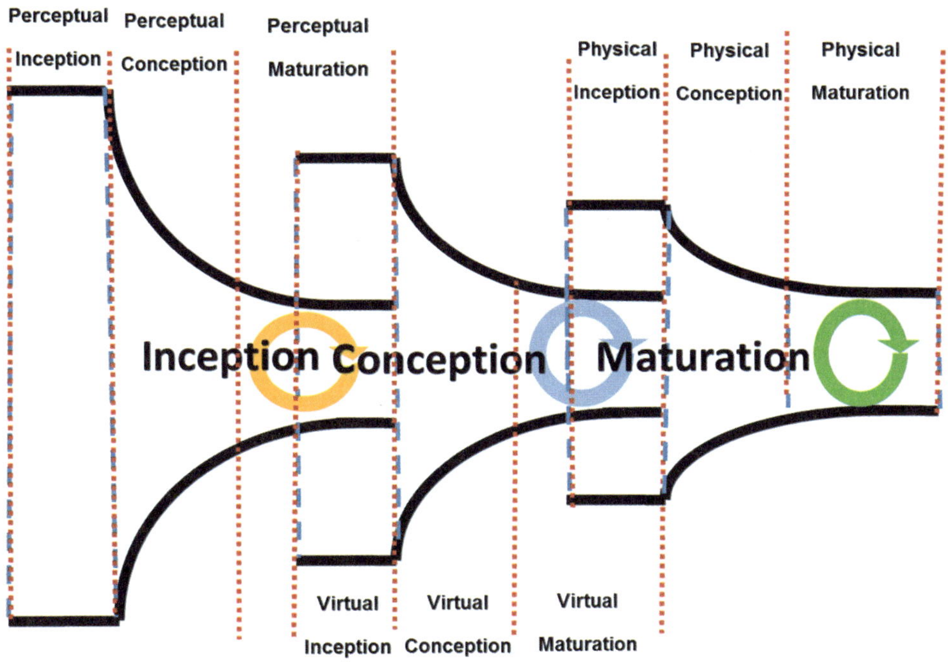

Figure 9.2 Three integrated phases of the new product realization process.

For craft production, the realization process aim is the design, development, and physical build of the desired product. For mass production, however, the realization process aim goes beyond the product to the design, development, and build of the production process as well. Traditionally, for new mass-produced products, the product must be fully defined before the production process can be realized. However, not accounting for process capabilities, limitations, and constraints during product definition can cause many (and possibly) costly iterations. To avoid the time and cost of such iterations, the product and production process should be simultaneously realized from inception to the end of physical realization.

As shown in Figure 9.3, during physical realization, specified production process facilities and tooling are procured, installed, and validated, while specified mass produced product subsystems and components are approved, assembled, and the full product is physically validated. The arrows in Figure 9.3 indicate the exchange of product and process information during every physical realization phase.

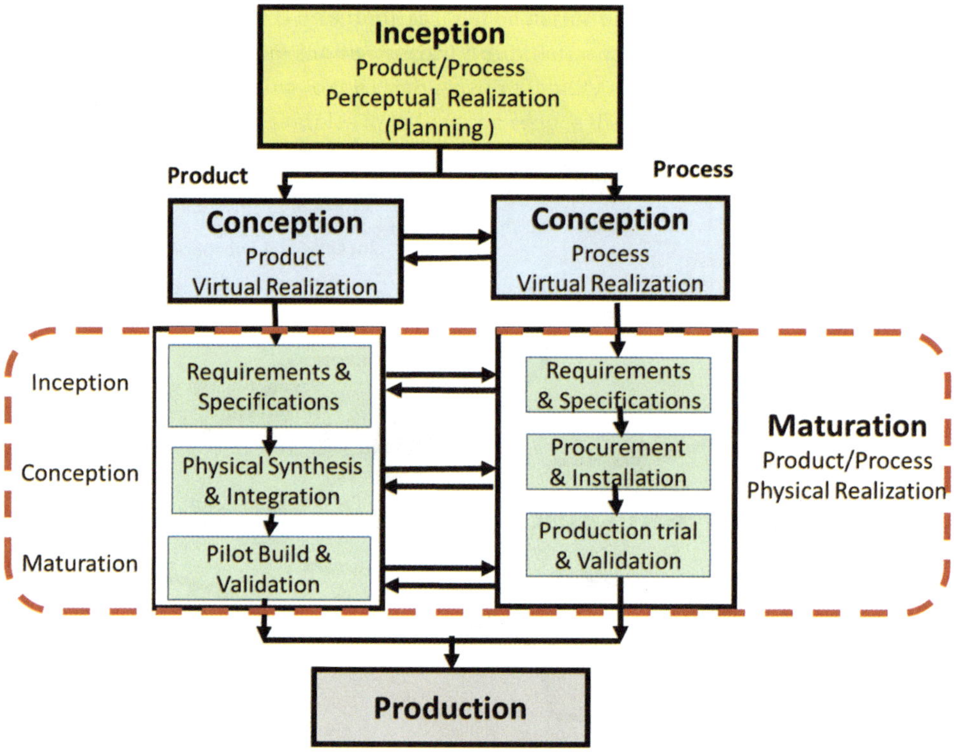

Figure 9.3 The joint physical realization of the product and production process.

Due to the linkage between the product and process characteristics, attempting to optimize each independently may be hard to achieve. For instance, product manufacturability is the outcome of process utility. Process utility, in turn, is a function of process architecture, execution, health, and fit characteristics and attributes, as shown in Figure 9.4.

Product manufacturability can constrain product architecture, disturb product health, and ultimately impede product utility. Subsequently, product quality and affordability, as fit characteristics, could be diminished. Since quality control is a process fit characteristic, it is clear that quality control strongly shapes product quality. Similarly, profitability as a process fit characteristic is strongly linked to affordability as a product fit characteristic. Accordingly, the main drive for quality planning and cross-functional team activities is to concurrently achieve the product key fit characteristic of quality and the process key fit characteristic of quality control, shown in Figure 9.4. Therefore, product quality and process quality control activities must be conducted simultaneously by the cross-functional quality planning team.

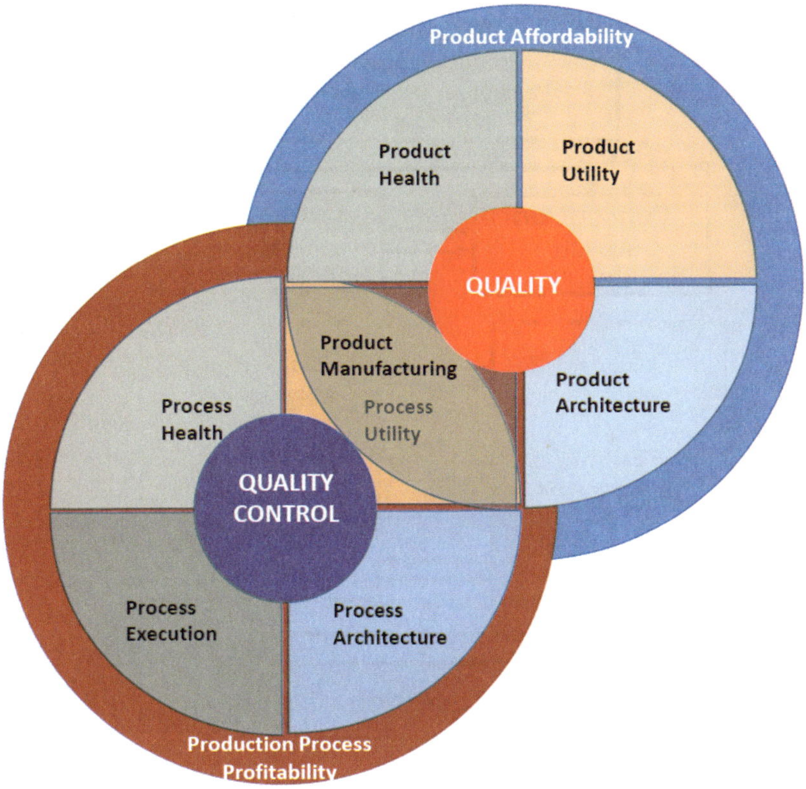

Figure 9.4 Connections between the process and product characteristics.

During physical realization, the main goal of quality planning activities is to assure that the product/process are physically integrated, developed, and validated as close as possible to the established targets. Therefore, not adding any development gaps and closing all variation gaps during physical realization should be the key pursuit of quality planning activities. Not adding any development gaps and closing all variation gaps, however, must start with the realization process inception. For this reason, advanced product quality planning activities must be well-integrated with all product realization phases and activities.

For achieving product and process targets and closing development and variation gaps during physical realization, the product quality cycle shown in Figure 9.5 could be followed. As indicated, during product/process physical maturation, iterations between development/validation and corrections/improvements are continued until all validation issues are resolved and desired characteristics are achieved.

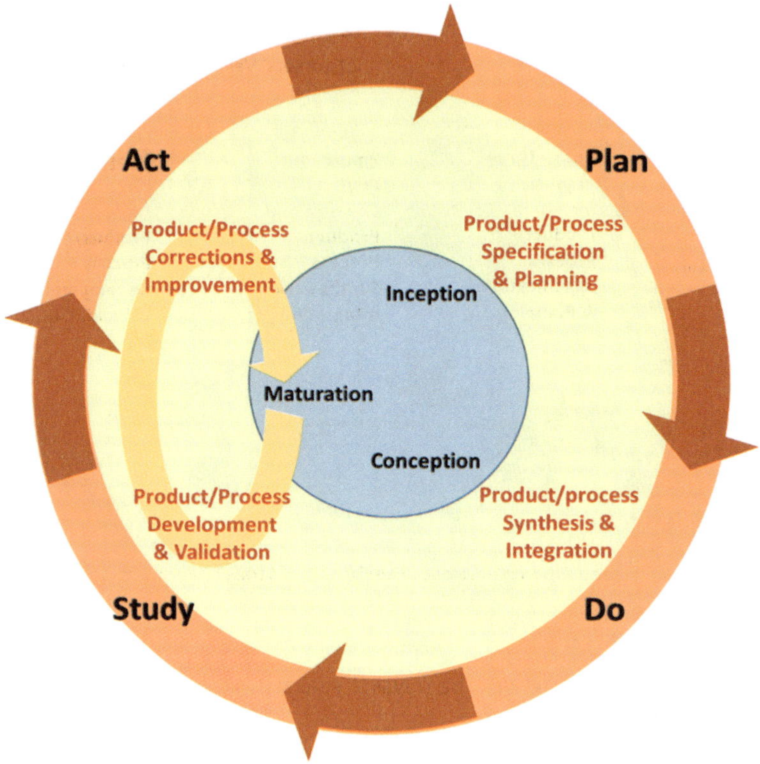

Figure 9.5 The physical realization quality planning cycle.

9.3 Physical Realization Model

Similar to the perceptual realization and virtual realization models, the physical realization model, shown in Figure 9.6, transforms the validated virtual product and process concepts to a mass produced product and production process. By conducting physical specification simultaneously with validation activities of virtual maturation, the physical specification becomes part of phase three. Starting with the specified product/process, phase four transforms the specified product/process to production concept through physical synthesis and integrations. Finally, the physical realization is concluded with phase five, during which the physical maturation of the product/process is completed through development/validation and corrections/improvements iterations. The physically developed product is validated through components validation tests, pilot validation tests, customers, and quality reviews. The process is developed and validated through trial runs and at rate preproduction run.

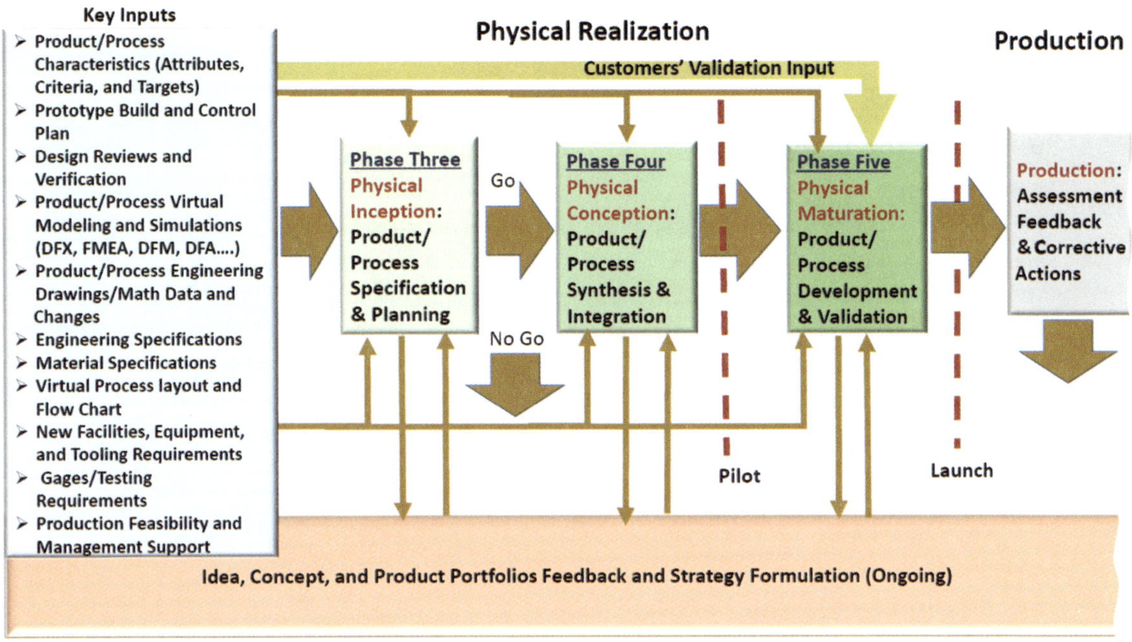

Figure 9.6 The physical realization model.

Iterations inside and between physical inception, conception, and maturation activities are possible if needed. However, the majority of iterations are usually needed during the maturation phase, between development/validation and corrections/improvements, as shown in Figure 9.5, until desired outcomes are realized. For example, during physical conception, any needed specification adjustments or changes could be easily done without the need to go back to inception. Similarly, during physical maturation, any needed adjustments or changes resulting from validation tests can be easily addressed through further development without having to go back to conception. In the following, the inception, conception, and maturation activities of the physical realization model, shown in Figure 9.6, in addition to the key inputs and outputs, will be outlined.

9.4 Physical Realization Inputs

As shown in Figure 9.6, the key inputs to physical realization from virtual realization are as follows [9-2], [9-3], [9-4], [9-5], [9-6], [9-7], [9-8], [9-9].

9.4.1 Product/Process Characteristics (Attributes, Criteria, and Targets)

The voice of the customer is specified and detailed in the matrix of product characteristics, attributes, criteria, and targets. In addition, the matrix of process characteristics, attributes, criteria, and targets enables the production of the customers' desired product.

Therefore, communicating these product/process characteristics, particularly in transition between realization domains, can provide a clear picture of the desired characteristics and the deciphered targets.

9.4.2 Prototype Build and Control Plan

Prototype build is the physical validation of the virtual realization or product design. Even though some of the off-the-shelf parts and subsystems may be used, prototypes are mainly craft produced and, therefore, costly to build. Vehicle prototypes are concept vehicles representing the physical proof of the full vehicle concept. Therefore, tests, assessments, and reviews including comprehensive customers' reviews of the concept vehicles are vital to assure that the perceptual and design portion of the development quality gaps are closed, or within acceptable limits. In addition, these prototype tests, assessments, and reviews can provide the following:

- Feedback on customers' perceptual summative product fit characteristics. This feedback is essential to close the gaps between desired and realized product characteristics, attributes, criteria, and targets.
- Feedback on form and function characteristics. This is necessary to address design issues and close development gaps.
- Insights on production process feasibility and requirements.
- Evaluation of timing, financial, marketing, and business goals. Evaluations include product affordability, investments, and return on investments.

Control plans provide detailed records and checklists of the architectural, materials, and functional tests to be performed during the build. Since prototypes are the initial physical inceptions, their build represents the initial physical build simulation. Accordingly, prototype build experience and control plan provide the initial model and simulation for pilot build, and eventually mass production.

9.4.3 Design Reviews and Verification

Design reviews must be planned, conducted by cross functional team, and formally documented. The main objective of design reviews is to track issues and monitor progress of product and process virtual realization. During design reviews the results of virtual inception, conception, and validation issues are usually discussed. Subsequently, design verification is the culmination of virtual development and validation iterations to ensure that the virtually realized product and process satisfy the constraints and meet the specified targets.

9.4.4 Product/Process Virtual Modeling and Simulations (DFX, FMEA, DFM, DFA, ...)

Virtual modeling and simulations are the means for design for X (DFX). X represents any target or specification necessary for achieving desired product and process criteria, attributes, and characteristics. For example, DFX activities include design for safety, durability, reliability, manufacturability, and assembly.

Product or process performance failure would occur if a target or specification is not met or a criterion was not accounted for. Failures because of unmet targets or specifications can be easily recognized. However, failures because of unaccounted-for criteria are hard to detect. Accordingly, design FMEA (DFMEA) is performed to account for potential failures because of unaccounted-for criteria. As a well-developed and disciplined methodology, FMEA is one of the key quality tools [9-2].

9.4.5 Virtually Realized and Validated Product/Process Engineering Drawings/Math Data and Changes

Math data and engineering drawings are the communications of the virtually realized and validated product/process. Accordingly, product/process math data and engineering drawings are the accumulated records of the perceptual and virtual realizations. Therefore, these records must be continuously reviewed and updated to reflect the latest changes, corrections, and improvements. Adequate definitions, dimensions, and control information should be present for every part and process station. Information should be compatible with manufacturing and industry standards, and communicable with all participants in the supply chain.

9.4.6 Engineering Specifications

Meeting product and process technical specifications is vital for achieving desired criteria, attributes, and characteristics. Therefore, key vehicle, components, and subsystems specifications must be included in the control plan. In addition, production process specifications, including suppliers' processes, should be well-detailed, virtually validated, and sufficient [9-3], [9-4], [9-5], [9-6], [9-7], [9-8].

9.4.7 Material Specifications

For virtual and physical modeling and simulations, material specifications and properties data must be continuously reviewed and updated to reflect the latest selections. In addition, materials with specific properties, handling, storage, and environmental requirements must be included in the control plan [9-9].

9.4.8 Virtual Process Layout and Flowchart

With the established specifications, the virtually realized production process layout and material flow can be optimized. The virtual process model and virtual simulations can be utilized before physical realization to optimize and finalize plants, material flow, and stations layout. Using established process targets for key attributes and criteria such as capability, productivity, and profitability, an optimized plant layout and flow can be conceptualized, virtually realized, and specified for physical realization.

9.4.9 New Facilities, Equipment, and Tooling Requirements

New facilities, expansions, equipment, and tooling to meet production capacity and timing could be determined during virtual modeling and simulations of process layout and flow. Timing and resource allocations for these new additions or expansions should be documented and communicated for accurate feasibility assessment.

9.4.10 Gauges/Testing Requirements

Before the start of physical realization, requirements for gauges and testing equipment must be established. Accordingly, timing and resource allocations for procurement and verification of these gauges and testing equipment should be identified, documented, and communicated for accurate feasibility assessment.

9.4.11 Production Feasibility and Management Support

Production feasibility must be conducted by the cross-functional team, and proper documents should be provided for management support. By reviewing the virtually realized product and process, the cross-functional team must determine the following:

- Virtually realized product meets customers' expectations, and quality gaps are closed or minimized.
- The virtually realized process is capable of producing the product as specified.
- All timing, financial, marketing, and business goals could be met.

9.5 Physical Inception

As the beginning of physical realization, physical inception is the interface with virtual realization, as shown in Figure 9.7. Physical inception could be simultaneously performed and integrated with the validation activities of virtual maturation unless mass production was not intended or deliberately suspended. As discussed in the previous chapter, virtual validation or concept proofing begins with prototype build and customers' reviews. Guided by feedback from customers' reviews and virtual validation results, in addition to portfolios' input, the vehicle concept can be enriched, and all necessary changes and improvements can be made during virtual maturation. As the virtual maturation activities are tapered down, the physical inception activities are ramped-up, as shown in Figure 9.7.

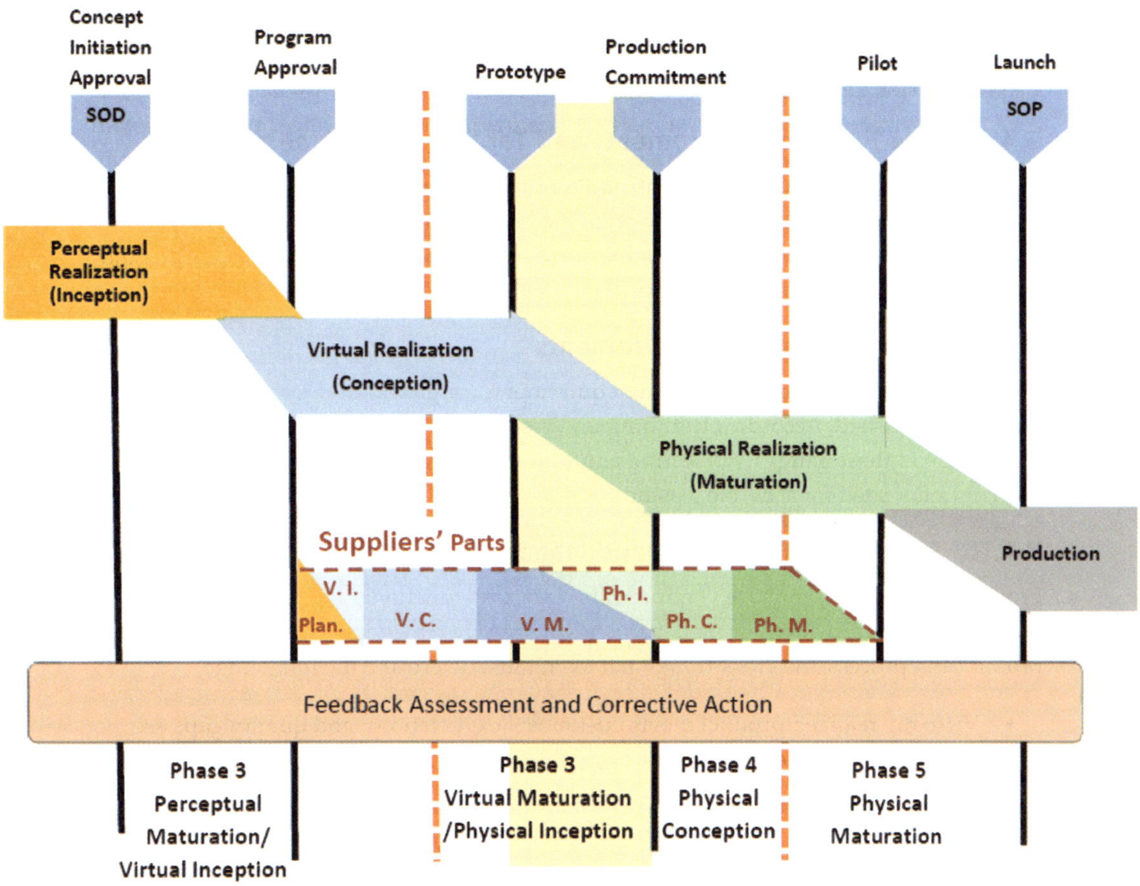

Figure 9.7 The physical inception phase of the new product realization process.

9.5.1 Product Form

Virtual realization and the proof of concept are completed once design verification, customers' evaluation, and all necessary corrections and improvements are completed. All changes and improvements as well as design verifications of all components must be completed during the virtual maturation iterations. As validation activities are being tapered at the end of virtual realization, the form of all components, subsystems, and full vehicle are physically specified through the physical inception activities. Bill of materials and all product specifications should be finalized before the beginning of physical conception.

9.5.2 Product Function

Planning physical validation testing and specifying test requirements for components, subsystems, and full vehicle must be completed during physical inception. By conducting physical inception activities simultaneously with virtual validation, the learnings from prototype build and validation activities can be utilized in specifying and planning key components testing and validation requirements, including all supplied parts.

9.5.3 Process Form

During virtual realization activities, phase lags between product virtual realization and production process virtual realization should be reduced as much as possible. Components' and subsystems' virtual realizations should not be considered complete during design verification until the production and assembly process design, modeling, and simulations are completed and verified as well. In some cases, production simulations could even lead component design, especially when the process limitations are the most dominant constraints. Additionally, after the initial design of a component and its production process are achieved, a joint optimization procedures could be performed with consideration of all objectives and constraints from both sides. At any rate, process design, optimization, and verification must be completed during process virtual maturation before the end of virtual realization. Meanwhile, during product/process physical inception, the integrated suppliers' parts/processes physical conception must be concluded before the beginning of full product/process physical conception.

9.5.4 Process Function

Based on established process attributes, criteria and targets discussed in Chapter 3, new facilities, equipment, tooling, and gauges are identified and specified during virtual realization. Subsequently, plants and stations specifications should be finalized during physical inception. For the production process function, using the finalized component technical specification, the manufacturing/purchase operation and subsequently production/quality controls can be derived. Building on the virtual realization house of quality discussed in Chapter 8, deriving the technical specifications of the manufacturing/purchase operations and production/quality controls can be facilitated, as shown in Figure 9.8.

With the final specifications during physical inception, the virtually realized and validated production process form and function can be enriched and optimized using virtual modeling and simulations. Accordingly, all process plants, stations layout, and material flow can be finalized before the physical build. To illustrate, the following example is considered.

9.5.4.1 Example 9.1: Production process layout and material flow can be optimized and finalized during physical inception with virtual modeling and simulations, as illustrated by Figure 9.9 [9-10].

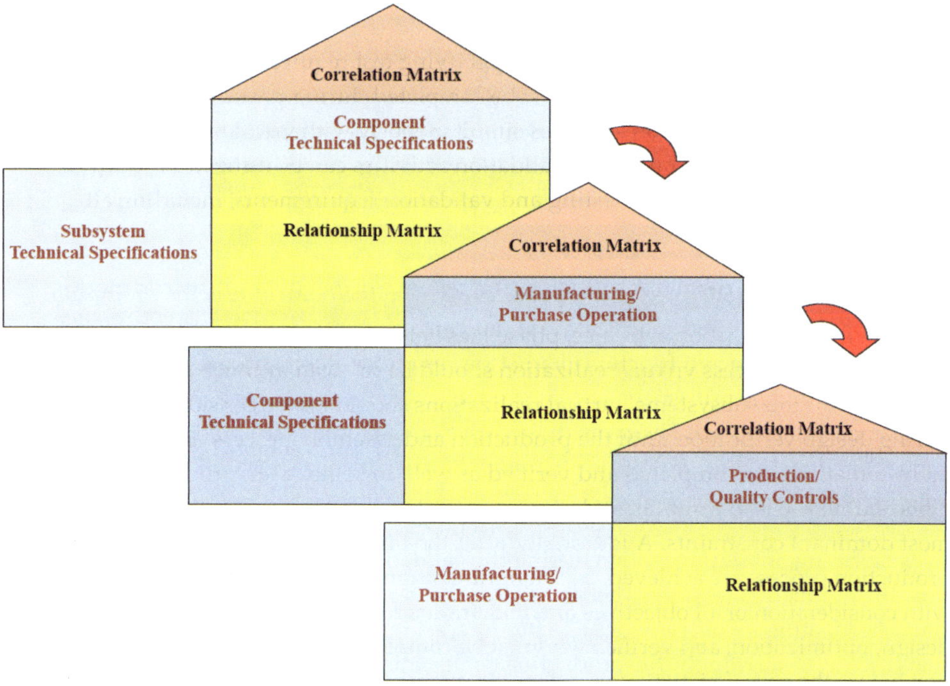

Figure 9.8 House of quality for manufacturing/purchase operations and production/quality controls.

Figure 9.9 Virtual modeling and simulation of production process flow [9-10]. (Chesky/shutterstock.com)

9.5.5 Product/Process Fit

With the availability of virtual modeling and simulation tools, the optimization process for desired criteria can be greatly facilitated. Therefore, building on the achieved virtual realization process capabilities, the enriched and specified physical process preliminary capability studies can be performed during physical inception. Preliminary process capability studies can be performed using the 'capability indices' discussed in Chapter 5 [9-11], [9-12], [9-13], [9-14], [9-15].

To ensure product, subsystems, and parts integrity at the point of use, the preliminary packaging standards and specification should also be developed. Packaging should protect product form and function characteristics under all conditions. Also, to assure that the required measurement system is established, the measurement system analysis plan should be finalized. The measurement system analysis plan should confirm that gauges of critical characteristics are calibrated for accuracy, repeatability, reliability, and validity [9-16], [9-17].

By performing physical inception during the validation activities of the virtual realization, the production feasibility becomes clearer and most potential production issues could be identified. This provides a better forecast for capital investments and production capacity. Accordingly, the production commitment decision can be facilitated and enabled. After production commitment is made, the physical conception and maturation could proceed until product launch.

By the end of physical inception, preliminary prelaunch control plans should be developed. In general, control plans are developed to identify product/process additional controls to be implemented. These additional controls are usually needed to contain potential nonconformance before full production and until production process is validated. The rest of the virtual realization inputs are used to guide product and process physical conception. Consequently, the output list of physical inception, as shown in Figure 9.10, include the following:

- new facilities, equipment, tooling, and gauges specifications
- physical process, flow chart, and floor plan layout specification
- preliminary packaging standards and specification
- engineering specifications
- material specifications
- product/process virtual modeling and simulations (DFX, FMEA, DFM, DFA, ...)

- preliminary physical process capability study plan
- preliminary measurement systems analysis plan
- preliminary pilot build and prelaunch control plans
- preliminary production part approval plan (PPAP)
- production commitment and management support

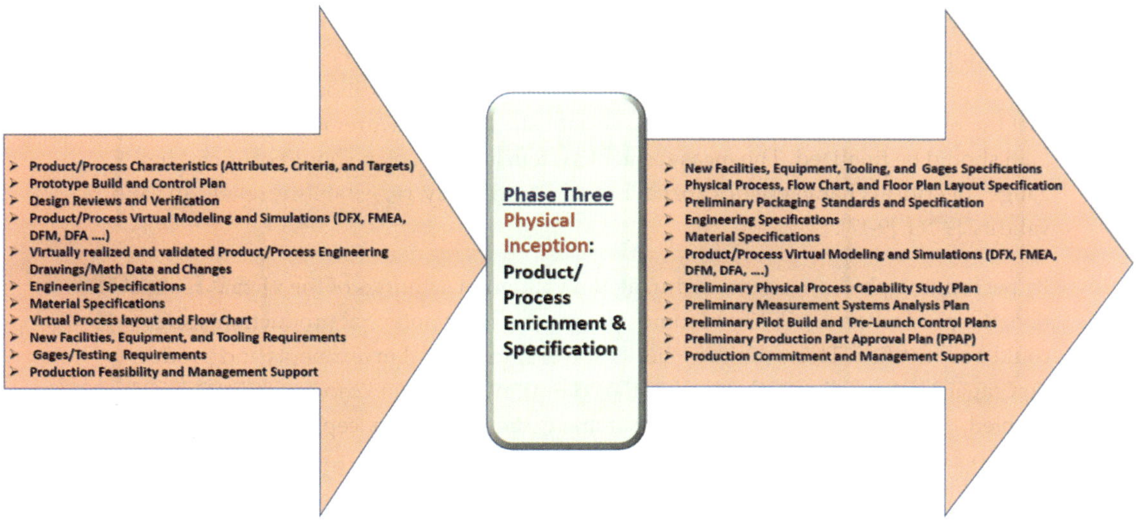

Figure 9.10 Physical inception inputs and outputs.

9.6 Physical Conception

The physical conception, or phase four, is the physical synthesis and integration of the specified product/process. During physical conception, the specified product/process is transformed into a physical production concept. For this reason, physical conception starts with production commitment, as shown in Figure 9.11. Consequently, physical procurement, integration, installation, and build of production plants and facilities are initiated. Maturation of supplied parts must start at physical conception, and all validation testing/improvements must be completed before the pilot build. Physical samples of supplied parts must be available for physical architectural integration.

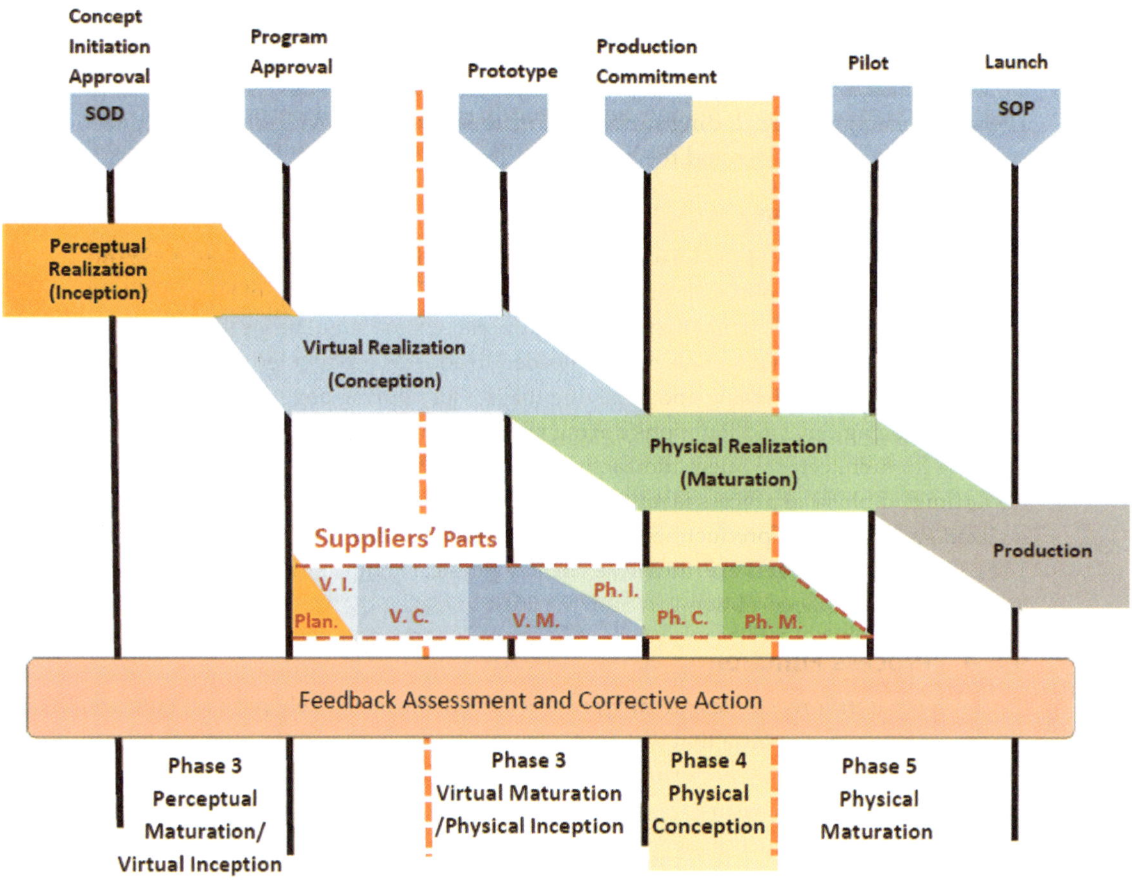

Figure 9.11 Physical conception phase of the new product realization process.

9.6.1 Product Form

Product physical form integration begins with architectural integration of production intent parts. Prior to suppliers' full production geometry, physical appearance, and specifications of supplied parts must be met. To ensure compliance with requirements and minimize risks, inspections of a supplied sample of parts and review of supplier process documentations are conducted. Accordingly, suppliers must provide sample parts and the required documentation following the production part approval process (PPAP) to obtain part approval [9-18].

9.6.2 Product Function

During physical conception, part FMEA must be performed to account for potential failures because of physical build issues and other unconsidered criteria [9-19]. FMEA should be conducted with production intent parts including supplied parts. In addition to FMEA, sensitivity analyses, optimization results, and design of experiments performed during virtual realization can provide better understanding of the product performance and failure modes [9-20], [9-21], [9-22], [9-23], [9-24]. Some of these failure modes can also be discovered during testing under normal and abusive conditions. Therefore, testing parts including supplied parts to failure can provide deeper insights into potential failure modes and their causes.

9.6.3 Process Form

The physical realization of the virtually realized and optimized process layout and flow begins with the start of physical conception. Depending on the validity of the simulations performed during virtual realization, additional virtual simulations using the already developed virtual models may be needed to address product/process physical integration issues. Addressing and resolving these issues during physical conception is vital for reducing time and iterations during physical maturation. Accordingly, the main objective for architectural integration activities during physical conception is to achieve the optimized physical process layout and flow. Procuring and assembling the virtually realized and optimized production systems and stations to meet process specifications and quality control targets are, therefore, the key physical architectural integration activities during physical conception.

9.6.4 Process Function

Assuming that the process design targets are achieved, the most likely physical failures will occur because of criteria that were not accounted for. Consequently, process physical FMEA (PFMEA) should be performed to account for potential failures due to unconsidered criteria. PFMEA assures process reliability, durability, and robustness during physical build and production operation. Therefore, during physical realization, PFMEA must be performed to account for potential failures because of physical build issues and other unconsidered criteria. Some of these failure modes can be discovered during trial runs of different stations under normal and extreme conditions.

9.6.5 Product/Process Fit

For quality planning, the pilot build control plan must be completed at the end of physical conception. Typically, the pilot build marks the start of physical validation of product/process, as shown in Figure 9.11. The rest of the physical inception inputs are used to guide product/process physical maturation. Consequently, the output list of physical conception as shown in Figure 9.12 include the following:

- process physical layout and flow integration
- process instructions
- process failure mode and effect analysis (PFMEA)
- packaging standards and specification
- characteristics matrix

- product/process quality system review
- pilot build and prelaunch control plans
- production part approval plan (PPAP)
- measurement systems analysis plan
- preliminary production process capability study plan
- management support

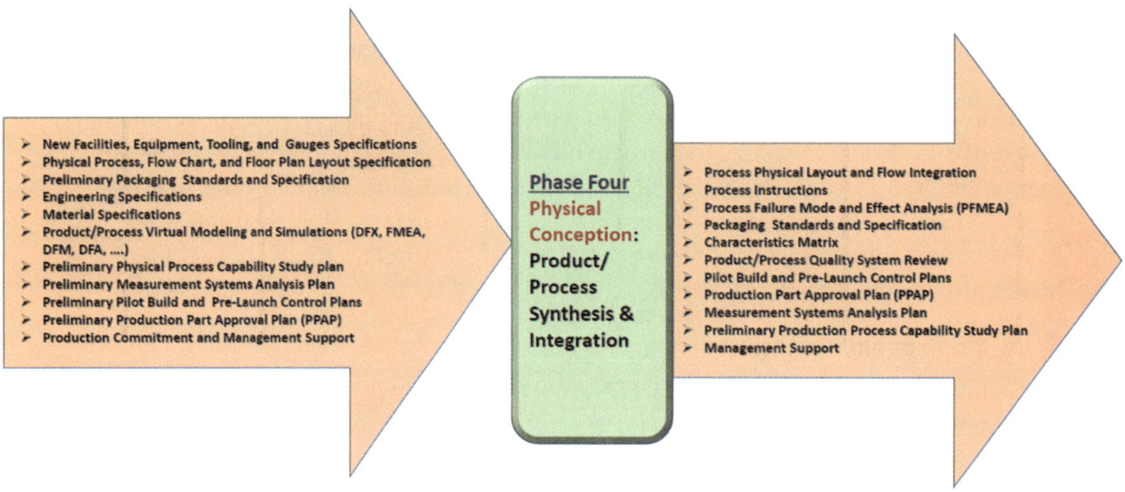

Figure 9.12 Physical conception inputs and outputs.

9.7 Physical Maturation

Product/process final development and validation are the main activities during physical maturation, as shown in Figure 9.13.

During the realization process and every realization phase, maturation develops, corrects, improves, and validates the conception before mapping to the next domain. While physical conception is the main product/process physical integration and build, physical maturation or phase five is the final development/validation, including the corrections/improvements before product launch. During physical maturation, the product/process is transformed into validated product and quality-controlled production-ready process.

For suppliers' parts as shown in Figure 9.13, the physical inception (Ph. I.) is ramped-up as virtual maturation (V. M.) is tapered down between prototype and production commitment. Consequently, physical conception (Ph. C.) starts with production commitment at the end of phase 3. Finally, physical maturation (Ph. M.) is tapered down between the end of phase 4 and pilot.

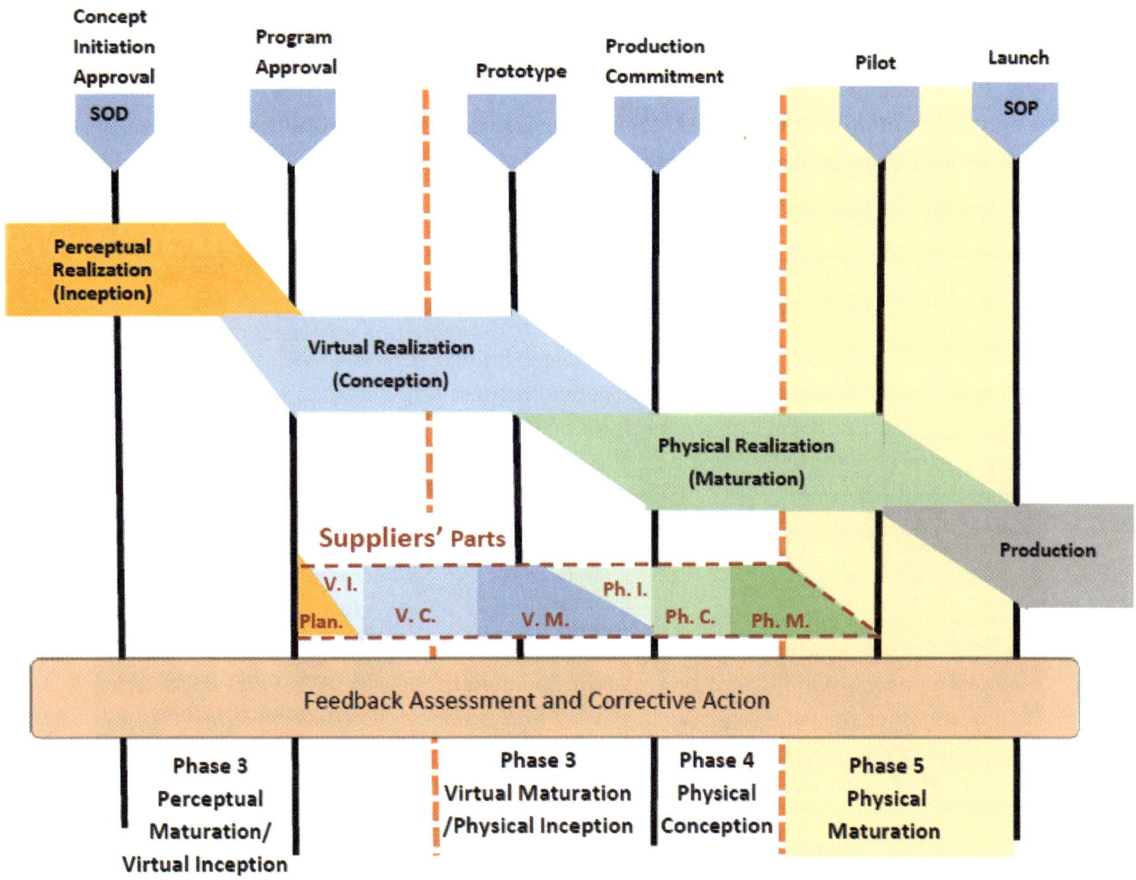

Figure 9.13 Physical maturation phase of the new product realization process.

9.7.1 Product Form

Product physical form maturation using production intent parts continues until pilot build. Production intent geometry, materials, and physical appearance of all subsystems, components and supplied parts must be in compliance with form specifications. All the remaining full-vehicle styling, packaging, and physical form integration issues must be addressed and resolved.

9.7.2 Product Function

The physical functional integration of production intent parts and subsystems usually start with physical laboratory testing. Depending on the availability and critical path, components and subsystems production intent laboratory testing, including suppliers' parts, could begin before physical maturation. Suppliers' parts laboratory testing must be completed before pilot build during suppliers' physical maturation. Consequently, all components and subsystems development laboratory testing should be completed before full-vehicle physical validation.

Laboratory tests conducted by suppliers must be performed according to the specified test condition [9-25], [9-26]. Test severity, duration, and the number of test samples must be specified according to the established reliability and confidence targets. In some cases, running test samples to failure can reduce the number of samples required for a specified reliability and confidence level. Testing incidents must be reported and addressed before full-vehicle physical validation activities. For physically simulated component testing, the following example is considered.

9.7.2.1 Example 9.2: To test the thermal effects on brake discs, the test rig shown in Figure 9.14 [9-27] is used.

Figure 9.14 Thermal testing of brake discs [9-27]. (Andrei Kholmov/shutterstock.com)

The pilot build and full-vehicle validation physical testing mark the beginning of the iterations between full vehicle development/validation and corrections/improvements activities of physical maturation. Validation testing of different characteristics and attributes should be performed according to full-vehicle technical specifications. Figure 9.15 [9-28] and Figure 9.16 [9-29], show examples of vehicle occupant protection during crash testing and vehicle handling testing.

Figure 9.15 Vehicle occupant protection during crash testing [9-28]. (Vereshchagin Dmitry/shutterstock.com)

Figure 9.16 Vehicle handling testing [9-29]. (istock.com/algre)

The number of validation test samples should be decided based on the targeted reliability and confidence level. It is important, however, to establish a strong correlation between the validation tests and the customers' physical realities. Validation tests are physical simulations for addressing failure due to geometry, materials, and production variations under controlled simulated loading and environmental conditions. Each customer's reality, however, presents a different loading and environmental condition that must be understood and covered by the controlled validation tests. Validation incidents must be documented using incident reports. Addressing these incidents during development/validation and corrections/improvements activities could be facilitated with virtual simulations using the updated virtual realization models and/or further accelerated physical laboratory tests.

9.7.3 Process Form

Supply chain integration must be finalized during physical development before the pilot build. All process architectural layouts, flows, stations, and logistics issues should also be addressed and resolved before the pilot build. Other operation issues related to process architecture should be addressed during process validation/development iterations and preproduction runs.

9.7.4 Process Function

All process function and performance issues from validation testing and production trial runs should be resolved during the physical maturation iterations. In addition, validation of measurement systems and packaging should be completed before product launch. Similar to product incidents, addressing process issues could be facilitated with matured virtual simulations using the updated virtual models before performing physical corrections and/or improvements.

9.7.4.1 Example 9.3: Physical realization of a body in white robotic station.

Using virtual modeling and simulation tools to optimize robotics set-up as shown in Figure 9.9, a body-in-white station similar to the one shown in Figure 9.17 [9-30] can be physically realized.

Figure 9.17 Physically realized robotic station [9-30]. (istock.com/Ric Aguiar)

9.7.5 Product/Process Fit

During physical maturation, all physical product and process FMEA should be completed. To confirm that all production intent parts meet specification, production part approval must be conducted. In addition, packaging evaluations are performed to ensure that parts are protected and can be utilized efficiently and safely at the point of delivery. By completing validation testing, packaging evaluation, suppliers' FMEA, and control plan production part approval can be completed.

For process physical validation, production trial run or run-at-rate is conducted. Run-at-rate or significant production run is also necessary for measurement systems analysis and initial process capability studies. Gauge repeatability and reproducibility should be validated by measuring the amount of variation in the measurement system arising from the measurement device and the operators taking the measurement. The measurement systems analysis is usually conducted using statistical tools such as analysis of variance ANOVA. As a living document, the control plan is updated to account for any process alterations. Therefore, any corrections or improvements must be accurately documented and included in the initial production control plan before quality planning sign-off. With management support and the signing of the product quality planning summary and the sign-off report by the quality planning cross-functional team, the product and process are ready for production launch. With the completion of the physical maturation activities, the list of outputs, as shown in Figure 9.18, include the following:

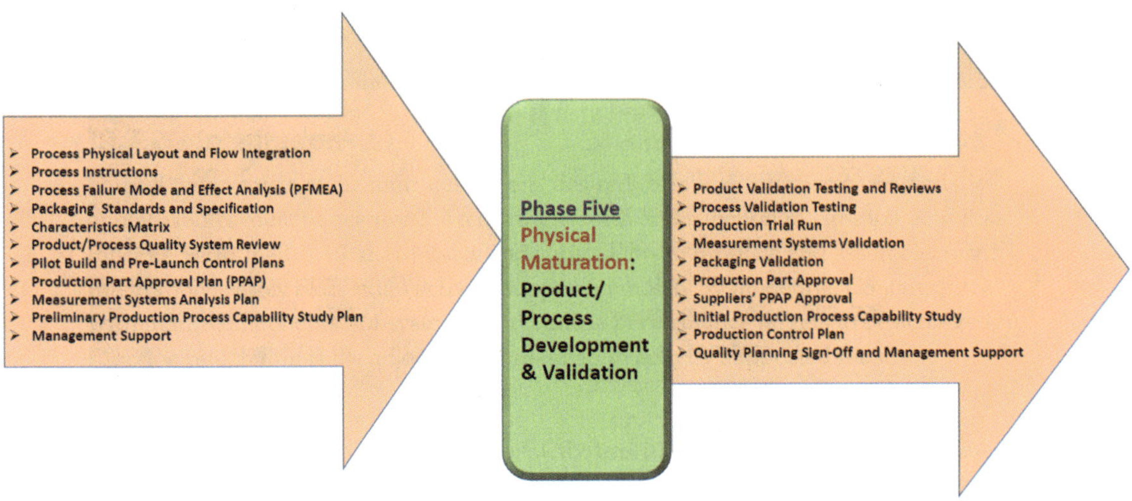

Figure 9.18 Physical maturation inputs and outputs.

- product validation testing and reviews
- process validation testing
- production trial run

- measurement systems validation
- packaging validation
- production part approval
- suppliers' PPAP approval
- initial production process capability study
- production control plan
- quality planning sign-off and management support

9.8 Production Assessment Feedback and Corrective Actions

Continual assessment and feedback after product launch are necessary for the duration of the production. Developing and implementing the proper corrective actions, by evaluating assessment feedback data and closing the loop, is the key to continuous improvement. Developed and implemented corrective actions should also be documented for improvement of the realization process of future products and production processes.

In addition to the product quality gaps discussed in Chapter 5, dealership and service experiences are integral parts of the customers' experience during product ownership. Therefore, customers' experiences with the product, the service, and the full ownership must be continually assessed. All available assessment data about customers' experiences—including surveys, audits, warranty data, and complaints—must be evaluated and any needed corrective actions must be taken. The assessments and evaluations outcomes of these customers' experiences throughout the ownership period should also be shared with the VOC activities for continuous improvements of new product preinception.

The continuous and timely delivery of quality vehicles and services depends on the continuous and timely delivery of quality parts and services throughout the entire supply chain. Because of the emotionally reactive nature of quality, the final full product quality is as good as the least quality of any part of the assembly. Therefore, products and services quality gaps of all supply chain participants should be continuously monitored and improved.

To guide corrective action efforts during production, the physical maturation outputs are provided as initial inputs to the production phase. Accordingly, the output list of production continual assessment feedback and corrective actions, as shown in Figure 9.19, include the following:

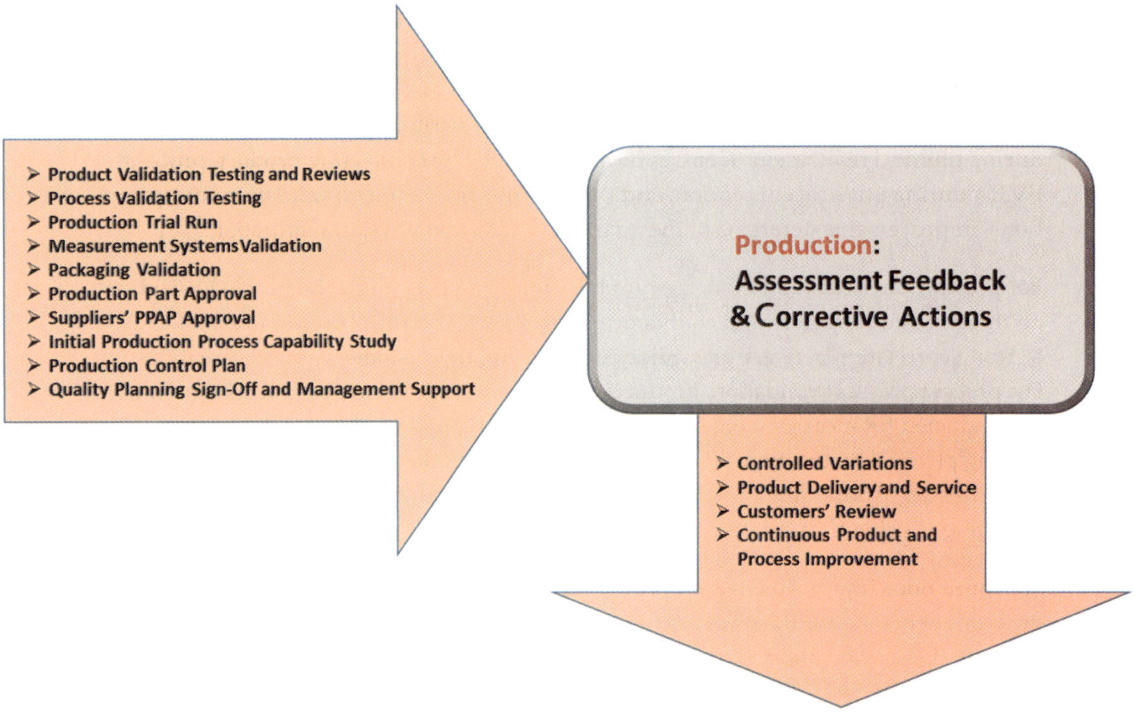

Figure 9.19 Production inputs and outputs.

- controlled variations
- product delivery and service
- customers' review
- continuous product and process improvement

9.9 Learning Outcomes from Vehicle Physical Realization

The desired learning outcomes from studying vehicle physical realization can be categorized as follows.

9.9.1 Knowledge

Physical realization is the preparation phase to mass production. Similar to any realization phase, the three identifiable physical realization subphases are physical inception, physical conception, and physical maturation. During physical inception or phase 3, the product and process virtual concept is enriched and further specified. The transformation of product and production process virtual realization into a product and production process physical concept through synthesis and integration begins with physical conception or phase 4. Development, validation, pilot build, and production approval of the product and production process are completed by product launch at the end of physical maturation or phase 5.

To achieve quality goals during physical realization, the product and production process virtual concept should be enriched, specified, integrated, developed, and validated as close as possible to the established targets based on the VOC. Closing all variation gaps in targeted form, function, and fit characteristics and attributes should be the main goal during physical realization. This can be accomplished by conducting product/process FMEA during physical conception, and through the development/validation, and corrections/improvements iterations of the quality cycle during physical maturation.

For product physical validation, the number of test samples should be decided based on the targeted reliability and confidence level. It is important, however, to establish a strong correlation between the validation tests and the customers' physical realities. For process physical validation, production trial run or significant production run must be conducted for measurement systems analysis and initial process capability studies. During physical maturation, all development and validation incident reports should be addressed. Addressing product or process incidents could be facilitated with virtual simulations using the updated virtual realization models.

For continuous improvement after product launch, continual assessment and feedback must be conducted for the duration of the production. Customers' experiences with the product, the service, and the full ownership must be continually assessed. Products and services quality gaps of all supply chain participants should be continuously monitored. To close the loop, all assessments, measurements, and feedback data must be evaluated for developing and implementing the proper corrective actions.

9.9.2 Skills

As the final realization phase, the physical realization is the connection between a concept and mass-produced vehicles. Accordingly, physical realization plays a pivotal role in achieving the desired product quality. Therefore, in addition to concurrent engineering and interpersonal skills, a high level of physical integration skills and attention to details are needed during physical realization. Without attention to details and striving for perfection during product and process physical realization, customers' desired form, function, and fit targets may not be achieved.

9.9.3 Paradigm/Attitude

Physical validation tests are physical simulations under controlled loading and environmental conditions for addressing failure due of geometry, materials, and production variations. Each customer's reality, however, presents a different loading and environmental condition that must be understood and covered by the controlled validation tests.

9.10 References

9-1. Kotha S. "Mass Customization: Implementing the Emerging Paradigm for Competitive Advantage," *Strategic Management Journal* 16(S1):21–42, 1995.

9-2. http://asq.org/learn-about-quality/tools-templates.html.

9-3. Wang, C. Julius. "Sample Size Determination of Bogey Tests Without Failures," *Quality and Reliability Engineering International* 7(1):35–38, 1991.

9-4. Kleyner, Andre, Shrikar Bhagath, Mauro Gasparini, Jeffrey Robinson, and Mark Bender. "Bayesian Techniques to Reduce the Sample Size in Automotive Electronics Attribute Testing." *Microelectronics Reliability* 37(6): 879–883, 1997.

9-5. Yadav, Om Prakash, Nanua Singh, and Parveen S. Goel. "Reliability Demonstration Test Planning: A Three Dimensional Consideration." *Reliability Engineering and System Safety* 91(8):882–893, 2006.

9-6. Kapur, K., and Michael P. 2014. *Reliability Engineering.* John Wiley & Sons: New Jersey.

9-7. Meeker, William Q., and Luis A. Escobar. 2014. *Statistical Methods for Reliability Data.* John Wiley & Sons: New Jersey.

9-8. Franceschini, Fiorenzo, and Domenico Maisano. "Standardisation of Quality and Reliability Tests in the Auto-Parts Industry: A Structured Approach Concerning Thermal Systems." *Total Quality Management and Business Excellence* Ahead of print.1–13, 2014.

9-9. http://erp.methode-eur.com/gifs/docs/GENERAL_GUIDE_AND_PROCS/AIAG%20%E2%80%93%20Advanced%20Product%20Quality%20Planning%20(APQP)%202nd%20Edition.pdf.

9-10. https://image.shutterstock.com/z/stock-photo-electric-vehicles-body-assembly-line-d-rendering-image-474075457.jpg.

9-11. Kotz, S., and Johnson, N. 1993. *Process Capability Indices.* Chapman & Hall: New York.

9-12. Chang, P., and Lu, K. *PCI Calculations for Any Shape Distribution With Percentile, QWTS,* pp. 110–114, 1994.

9-13. Kotz, S., and Lovelace, C. 1998. *Process Capability Indices in Theory and Practice.* Arnold: London.

9-14. Kotz, S., and Johnson, N. Process Capability Indices—A Review, 1992–2000. *Journal of Quality Technology* 34(1):2–53, 2002.

9-15. Ramanathan, T., Dharmadhikari, A., and Abraham, B. Nonparametric Capability Indices. *Economic Quality Control* 18(1):31–41, 2003.

9-16. Hetland, Gregory A. 1999. *Measurement Systems Analysis. Dimensioning and Tolerancing Handbook.* McGraw-Hill: New York.

9-17. Kooshan, F. "Implementation of Measurement System Analysis System (MSA): In the Piston Ring Company 'Case Study'," *International Journal of Science and Technology* 2(10):749–761, 2012.

9-18. http://b2b.methode-eur.com/gifs/docs/GENERAL_GUIDE_AND_PROCS/AIAG%20%E2%80%93%20Production%20Part%20Approval%20process%20(PPAP)%204th%20Edition.pdf.

9-19. Press, Dyadem. 2003. *Guidelines for Failure Mode and Effects Analysis (FMEA), for Automotive, Aerospace, and General Manufacturing Industries*. CRC Press: Boca Raton, Florida.

9-20. Cameron, T. M., L. Jordan, and M. E. M. El-Sayed. "Sensitivity of Structural Joint Stiffnesses With Respect to Beam Properties: A Hybrid Approach," *Computers and Structures* 63(6):1037–1041, 1997.

9-21. Cameron, T. M., A. C. Thirunavukarasu, and M. E. M. El-Sayed. "Optimization of Frame Structures with Flexible Joints." *Structural and Multidisciplinary Optimization* 19(3):204–213, 2000.

9-22. Fisher, Sir Ronald Aylmer. 1960. The design of experiments. 12(6). Oliver and Boyd: Edinburgh.

9-23. Montgomery, Douglas C. 1984. *Design and Analysis of Experiments*. Vol. 7. Wiley: New York.

9-24. Park, Gyung-Jin. *Analytic Methods for Design Practice*. 309–391, 2007. Springer: New York.

9-25. MTS. Component Testing Idea Book. TestLine Components/TestStand Solutions. 2007. Available from: MTS: Moscow. http://www.mts.com/ucm/groups/public/documents/library/dev_002115.pdf.

9-26. Panse, S. and Gosavi, S., "Integrated Structural Durability Test Cycle Development for a Car and its Components," SAE Technical Paper 2004-01-1654, 2004, doi:10.4271/2004-01-1654.

9-27. https://www.shutterstock.com/image-photo/brake-disc-thermal-test-car-discs-282652784?src=IPBAahebx9u5ScG8fzSqRg-1-4.

9-28. https://www.shutterstock.com/image-photo/image-mannequin-car-after-crashtest-59885752.

9-29. http://www.istockphoto.com/photo/car-industry-gm179584943.

9-30. http://www.istockphoto.com/photo/car-industry-gm157337497.

Chapter 10
Product Realization Enablers

"Desire is the key to motivation, but it's determination and commitment to an unrelenting pursuit of your goal—a commitment to excellence—that will enable you to attain the success you seek."

—Mario Andretti

10.1 Introduction

Sustained product realization success is essential for achieving targeted product quality, time to market, affordability, profitability, and, ultimately, continued organization growth. Therefore, enabling sustained product realization success must be a continuous organizational pursuit, especially in a highly competitive global market. Among the many product realization enablers to be considered, some of the key enablers can be identified by examining the following three basic elements of the quality domains discussed previously in Chapter 5:

- *Sender's Domain:* With the three enterprise elements of people, processes, and provisions
- *Product's Domain:* With the three sets of form, function, and fit characteristics
- *Receiver's Domain:* With the three perceptual functions of expectation (imaginative), evaluation (cognitive), and emotions (affective)

While successes in achieving desired product's domain, form, function, and fit characteristics are measured in the receiver's domain, the product realization key enablers are always in the sender's domain. In this chapter, brief discussions of these enablers' pertinent aspects will be conducted. The main objective is to demonstrate the role and importance of the three enterprise elements of people, processes, and provisions in empowering the continued success of the product, the realization process, and the organization.

10.2 People

People are the most significant enabler to the continued success of all aspects of the organization, including product realization. As discussed in Chapter 2, all product realizations start at the perceptual domains of people. In fact, all product realization activities and outcomes are conducted by people to serve people. Therefore, success of any organization will ultimately depend on how all internal and external people are viewed and treated. Consequently, valuing and respecting people should include all people, internal and external, starting with every member in the organization.

Regardless of their title or position, all members of the organization should feel valued and empowered, instead of being treated or considered as resources or commodities. For instance, having one part of the organization treat or consider the other part as mere resources creates two classes of people. These two classes are the decision-making class and the commodity class. Once the members of any group in the organization are commoditized their commitment becomes a mere compliance and eventually, their loyalty, trust, and motivation will wither. Subsequently, the organizational unity, harmony, and creativity will be diminished. Without the key characteristics of unity, harmony, and creativity organizational stability, productivity and innovation will not be sustained. Ultimately, the organizational ability to endure hard times in a competitive tempestuous market will vanish.

For successful empowerment, ongoing development of all people in the organization is vital for continued growth. To guide holistic development and growth, an example of human characteristics framework is illustrated in Figure 10.1.

As shown in Figure 10.1, human congenital and developmental characteristics represent the human equivalent to the form characteristics, while health and capabilities characteristics represent the human equivalent to the functional characteristics. The professional, relational, and personal human development capabilities can be developed through education, training, and practice. The underlining social abilities or sociability is the summative fit characteristics needed for dealing with others. The peak summative fit characteristics of refinement are the result of the continuous improvement of all other characteristics.

The three human innate abilities considered in Figure 10.1 are known as Bloom's development domains [10-1], [10-2], [10-3], [10-4], [10-5], [10-6], [10-7], [10-8], [10-9]:

- cognitive (thinking and comprehending)
- affective (values and emotional abilities)
- psychomotor (physical and movement abilities)

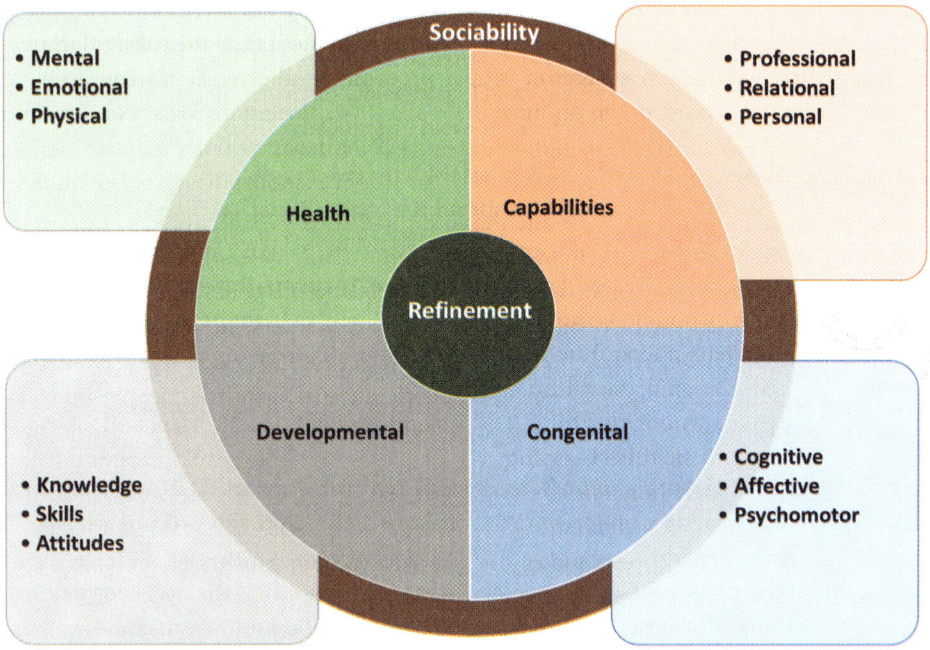

Figure 10.1 An example of a framework for human characteristics.

For the cognitive domain, Bloom's development levels are as follows:

1. *Knowledge:* Information gathering without necessarily understanding, using, or altering it
2. *Comprehension:* Understanding the gathered information without necessarily relating it to anything else
3. *Application:* Using general concept gained through comprehension to deal with an issue
4. *Analysis:* Disassembling something down into its fundamental elements
5. *Synthesis:* Creating something new by integrating different elements
6. *Evaluation:* Differentiating the subtle differences in objects or methods

Consider the following three main realization objectives discussed in Chapter 2:

- To understand reality (know it)
- To utilize reality (use it)
- To alter reality (improve it)

The connection between Bloom's cognitive development levels and the three realization objectives can be identified as illustrated in Figure 10.2 and Figure 10.3. As shown in these figures, Bloom's knowledge and comprehension correspond to understanding reality, which is usually associated with learning. Application and analysis correspond to utilizing reality, which is usually associated with problem solving. Synthesis and evaluation correspond to altering reality, which is usually associated with design for altering physical reality or research for altering perceptual reality.

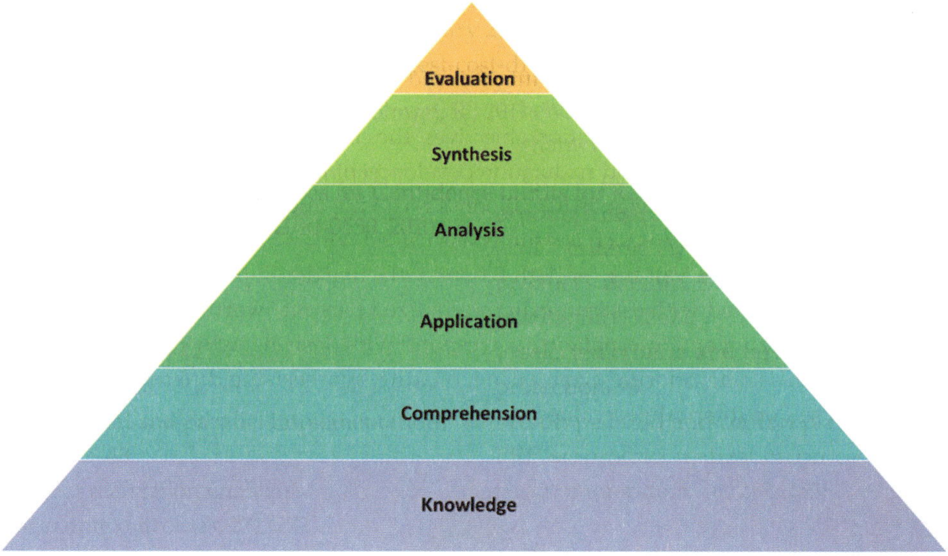

Figure 10.2 Bloom's cognitive development levels.

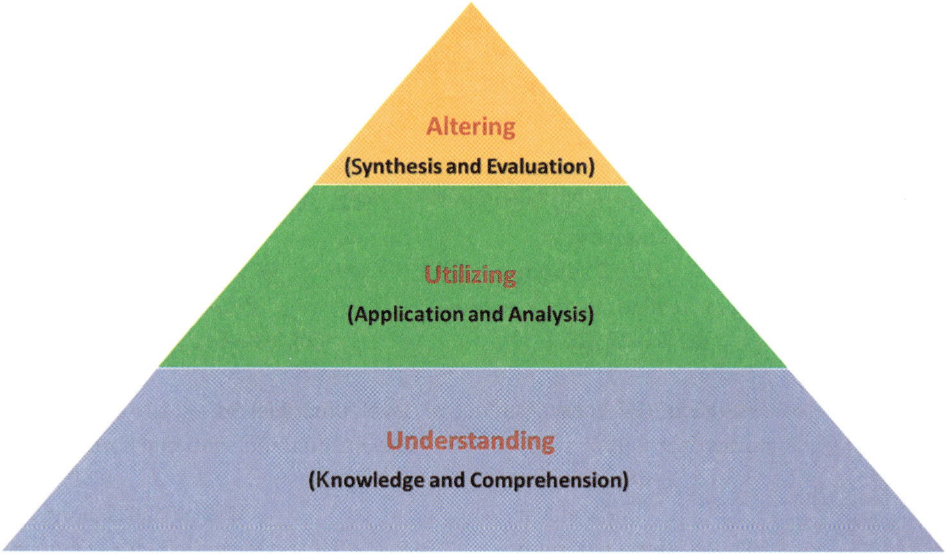

Figure 10.3 Realization's cognitive development levels.

For the affective domain, Bloom's development levels are as follows:

1. *Receiving:* Awareness and willingness to receive
2. *Responding:* Willingness and active participation in responding (motivation)
3. *Valuing:* Attaching different worth or value to a particular object or action
4. *Organizing:* Setting priorities, comparing, relating, and synthesizing different values
5. *Internalizing:* Behaving based on internalized value system

Considering the three main realization objectives, the correspondence between Bloom's affective development levels and realization development levels can be demonstrated as shown in Figure 10.4 and Figure 10.5. Bloom's receiving and responding correspond with engaging with reality, which is usually associated with willingness to receive and be an active participant. Valuing for both Bloom's affective development and realization objectives is associated with assigning an internalized personal worth to a specific object or action. Organizing and internalizing correspond with desiring, which is usually associated with setting priorities, comparing, relating, synthesizing different values, and behaving based on internalized values.

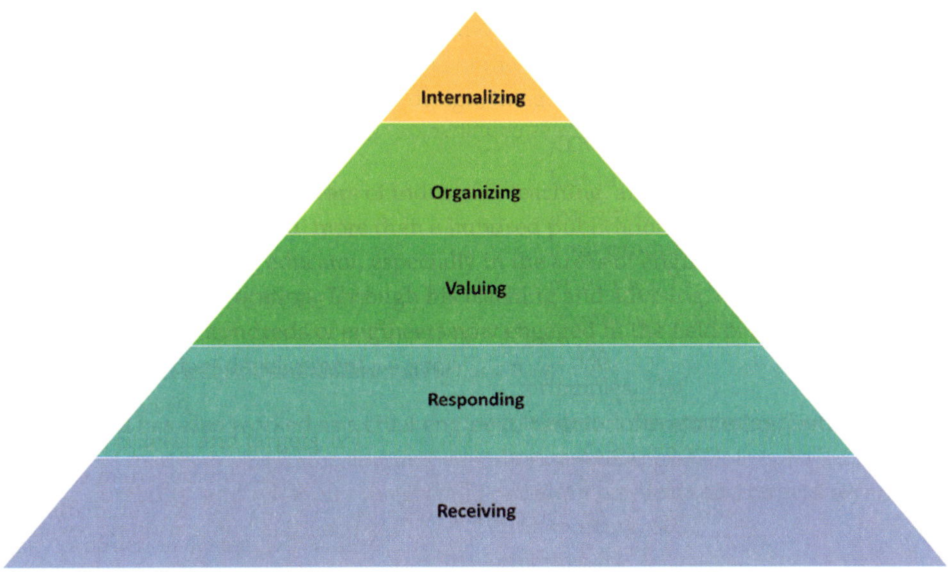

Figure 10.4 Bloom's affective development levels.

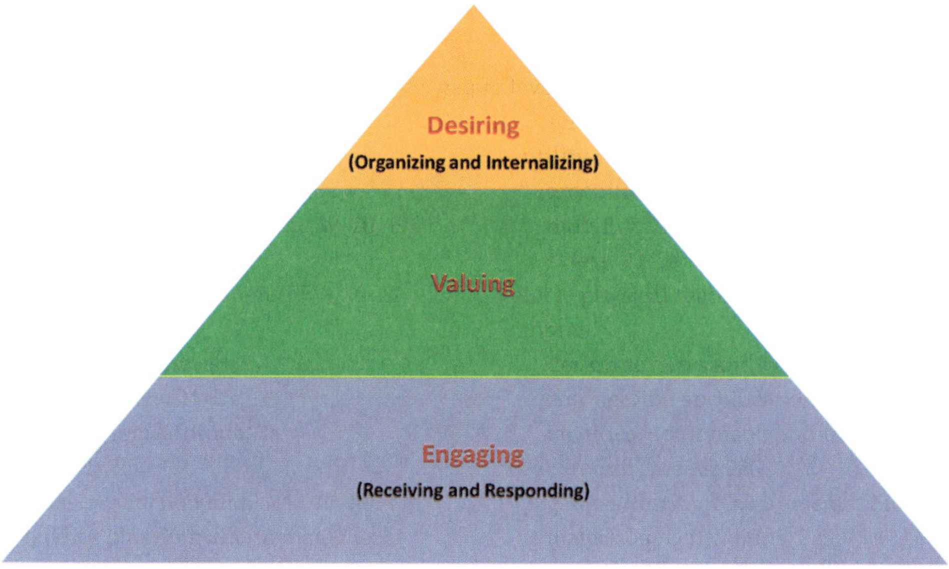

Figure 10.5 Realization's affective development levels.

Some of the connections between the affective development levels, cognitive development levels, and realization objectives can be identified from Figure 10.3 and Figure 10.5. For instance, being engaged stimulates the understanding of a reality. Also, valuing what a reality can offer motivates its utilization while desiring more from it inspires its alteration.

For the psychomotor domain, Bloom's development levels are as follows:

1. *Perception:* Guiding motor activity using sensory cues
2. *Set:* Getting ready to act mentally, physically, and emotionally
3. *Guided Response:* Starting to learn complex skills through imitation, and trial and error
4. *Mechanism:* Gaining confidence and proficiency in learning complex skills
5. *Complex Overt Response:* Performing complex movement skillfully
6. *Adaptation:* Modifying movement patterns to fit specific requirements
7. *Origination:* Creating new movement patterns

Similar to the cognitive and affective domains, considering the three main realization objectives, the correspondence between Bloom's psychomotor development levels can be demonstrated as shown in Figure 10.6 and Figure 10.7. Bloom's perception, set, and guided response correspond to trying in physical reality, which is usually associated with preparing mentally, emotionally, and physically using sensory cues to learn new skills through imitation and trial and error. Mechanism and complex overt response correspond to mastering the skills, which is associated with reaching the state of internalized physical realization, confidence, and proficiency to perform a new physical action skillfully. Adaptation and organization correspond with refining by altering the physical action, which is the highest level of psychomotor development.

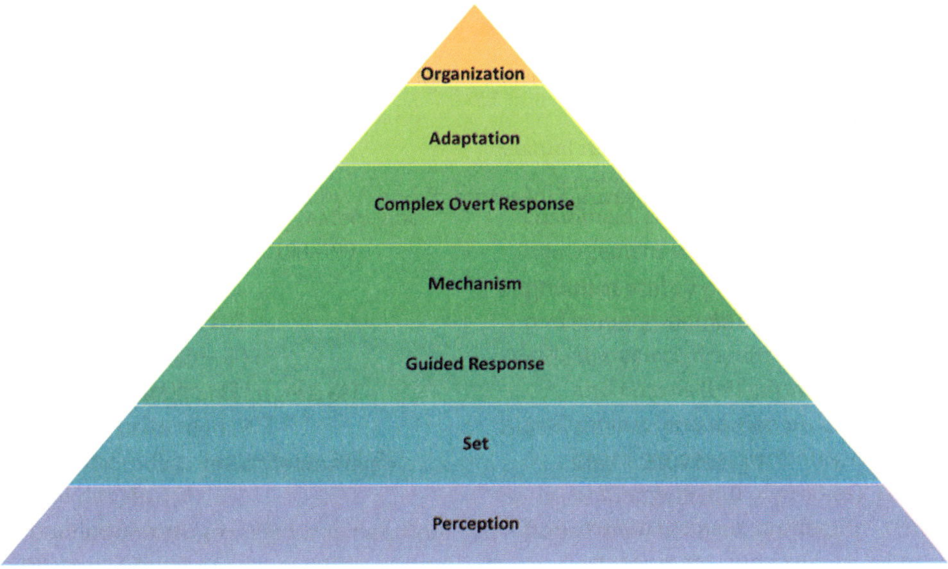

Figure 10.6 Bloom's psychomotor development levels.

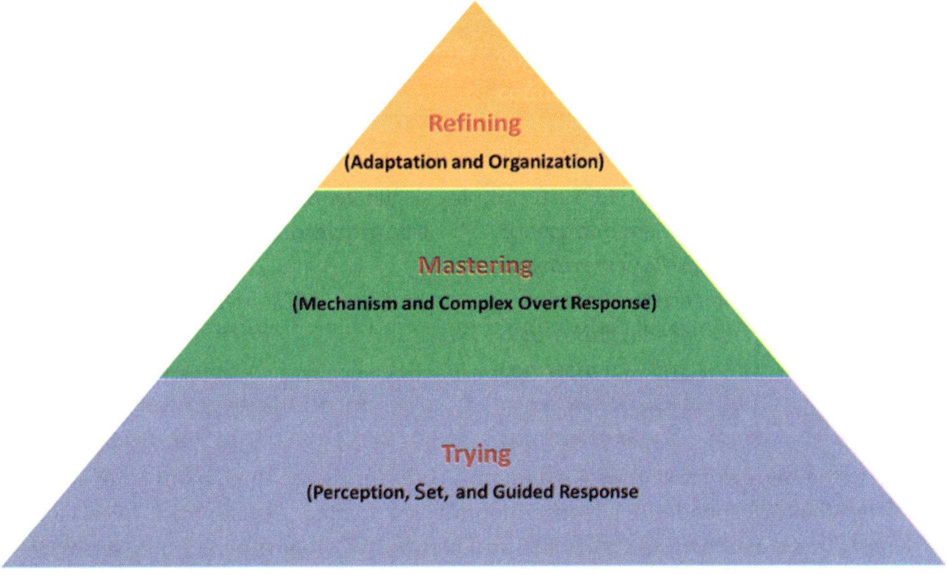

Figure 10.7 Realization's psychomotor development levels.

Some of the connections between the psychomotor, cognitive, and affective development levels can be identified from Figure 10.3, Figure 10.5, and Figure 10.7. For instance, to develop a psychomotor skill, one must cognitively understand the need for it and must be affectively engaged and willing to try. To master a psychomotor skill, one must value it and continuously utilize it. However, to refine a psychomotor skill, one must have the desire to reach new, uncharted, and ultimately altered cognitive and physical realizations.

The advantages of relating the development domains to the three main realization objectives are simplicity and the ability to identify, develop, and match people with the appropriate level of realization activities. For example, the cognitive level of utilizing must be reached for perceptual and virtual problem solving, while the psychomotor level of mastering would be necessary in physical problem solving. Similarly, the cognitive development level of altering must be reached for perceptual and virtual realizations, while the psychomotor level of refining would be necessary for physical realization. Accordingly, for sustained innovation in product realization, the altering and refining levels should be attained in cognitive and psychomotor domains, respectively.

Being mentally, spiritually, and physically healthy is essential for reaching the full potential of human performance capabilities, as shown in Figure 10.1. Knowledge, skills, and attitudes do not only affect continued performance, but they also affect the fit characteristics of sociability and refinement. Desiring, or at least having positive valuation in, the affective domain is necessary for human performance, sociability, and refinement. As stated by Mario Andretti, "desire is the key to motivation." In addition, commitment to excellence and the determination to achieve it are the results of acting on the internalized values in the affective domain. As continued by Mario Andretti, "but it's determination and commitment to an unrelenting pursuit of your goal—a commitment to excellence—that will enable you to attain the success you seek."

While cognitive and psychomotor development can be facilitated, achieved, and refined through education, training and practice, positive valuing, and ultimately desiring as affective traits are different. Due to their relational, emotional, and highly internalized nature, affective traits could only be fostered and nurtured. In any organizational setting, valuing people fosters mutual valuing in the affective domains of these people. Similarly, empowering people provide the sense of ownership and pride necessary for fostering the desire to succeed and excel. Therefore, as discussed earlier, all people in the organization should be respected and well treated to feel that they are valued and empowered for continued organizational unity, harmony, and creativity.

10.3 Process

For continued realization process success, some of the key enablers are connected to the process itself. These key enablers are associated with analysis, integration, and management. In the following analysis and its usage, integration teams and their structure, along with the vital role of management, will be discoursed as examples of key realization process enablers.

10.3.1 Analysis

As discussed in Chapter 2, analysis can be performed perceptually, virtually, and physically through modeling and simulations. Due to their many advantages, virtual modeling and simulations are frequently used during all three realization phases. In fact, analysis and virtual analysis are usually synonymous. Some examples of the virtual analysis advantages over perceptual analysis are the results of the virtual modeling and simulation abilities to allow the following:

- collective and comprehensive inputs
- visual and representative models
- complex and multifaceted simulations
- visual and repeatable outputs
- reliable and verifiable results

Similarly, some examples of the advantages of virtual analysis over physical analysis are the results of its ability to allow the following:

- different possible and alternative inputs
- faster and far less costly modeling
- faster and far less expensive simulations
- deeper insights for cause-and-effect detections
- multiobjective, multivariable optimization

Due to its advantages and abilities, virtual analyses are usually used during physical realization to enable addressing product and production process development and validation issues. As for perceptual realization, simplified virtual modeling and simulation tools, including simplified mathematical models, can be used for early product characterization during perceptual conception. To demonstrate the use of simplified virtual models, simulations, and optimization tools during perceptual conception, the following case study is considered [10-10].

10.3.1.1 Case Study 10.1: Preliminary Full-Vehicle Multiobjective Optimization [10-10]: A multiobjective optimization procedure was utilized to study different alternatives for optimized vehicle longitudinal acceleration and steady-state fuel economy performance. Fuel economy and five-speed manual transmission with engine at wide-open throttle, along with the ride and handling quality of the vehicle, were mathematically modeled. A multiobjective optimization algorithm was utilized to determine the full-vehicle design parameters, such as transmission gear ratios, tire radius, driving axle ratio, vehicle weight, wheel base, and position of the center of gravity.

In order to achieve targets within specified priorities, a composite objective function was utilized with two mathematically modeled objectives and two mathematically modeled constraints. The first objective is the 0–60 mph time, and the second objective is the steady-state miles per gallon (MPG). The two constraints considered are under-steer for vehicle stability and bounce frequency for ride and handling [10-10].

Table 10.1 [10-10] shows the initial values of the design case (a), and the optimization results for two different cases (b) and (c). For case (b), with both objectives having equal weight factors, the optimization process improved both objectives with reduction in vehicle weight. For case (c), by increasing the weight factor F_2 from 1 to 4, further fuel economy improvement was achieved in comparison with case (b).

Table 10.1 Multiobjective optimization results with different weight factors.

Design variables/ objectives	(a) Initial values	(b) Optimized values $F_1 = 1$ and $F_2 = 1$	(c) Optimized values $F_1 = 1$ and $F_2 = 4$
Second gear ratio	2.07	2.796	2.099
Third gear ratio	1.80	2.400	1.800
Fourth gear ratio	1.48	1.48	1.48
Wheel base (ft)	9.33	7.50	8.78
Position of CG (ft)	4.33	3.87	4.49
Tire radius (ft)	1.08	1.137	1.175
Driving axle ratio	3.89	3.50	3.15
Vehicle weight (pounds)	2400	2386.2	2391.8
Objective 1: 0–60 time (s)	8.37	7.53	7.79
Objective 2: miles per gallon	23.71	26.54	27.22

10.3.2 Integration

During new vehicle realization, customers' desired form and function characteristics and attributes must be achieved. Accordingly, form and function integration activities must be performed. With many parts and subsystems to produce and assemble, form integration is a key realization activity supported mostly by modeling. Concurrently, with many components, subsystems, and vehicle performance targets to be achieved, function integration is the adjoined key realization activity supported mostly by simulations.

To conduct form and function integration activities, integration teams must be formed. These teams can be classified as form integration teams for form areas (architecture and manufacturing) and function integration teams for function areas (utility and health). Integration activities for product fit characteristics such as quality and affordability should be led by a cross-functional team that incorporates representations from all integration areas.

The form integration area can be classified as architectural integration and manufacturability integration. Examples of the integration areas for product architecture include the following:

- exterior
- interior
- body
- chassis
- powertrain
- packaging

Examples of the integration areas for product manufacturability include the following:

- materials
- formability
- supply parts
- assembly
- treatment

Similarly, the function integration area can be classified as utility, health, and life-cycle integrations. Examples of the integration areas for product utility include the following:

- safety
- quality, reliability, and durability (QRD)
- drivability
- vehicle dynamics
- ride and handling
- noise, vibration, and harshness (NVH)
- thermal management
- electric/electronics
- harmony and human factors

Examples of the integration areas for product health and life cycle include the following:

- testing
- validation
- service
- maintenance

- repair
- recycling

Figure 10.8 illustrates an example of the different integration areas involved in product realization.

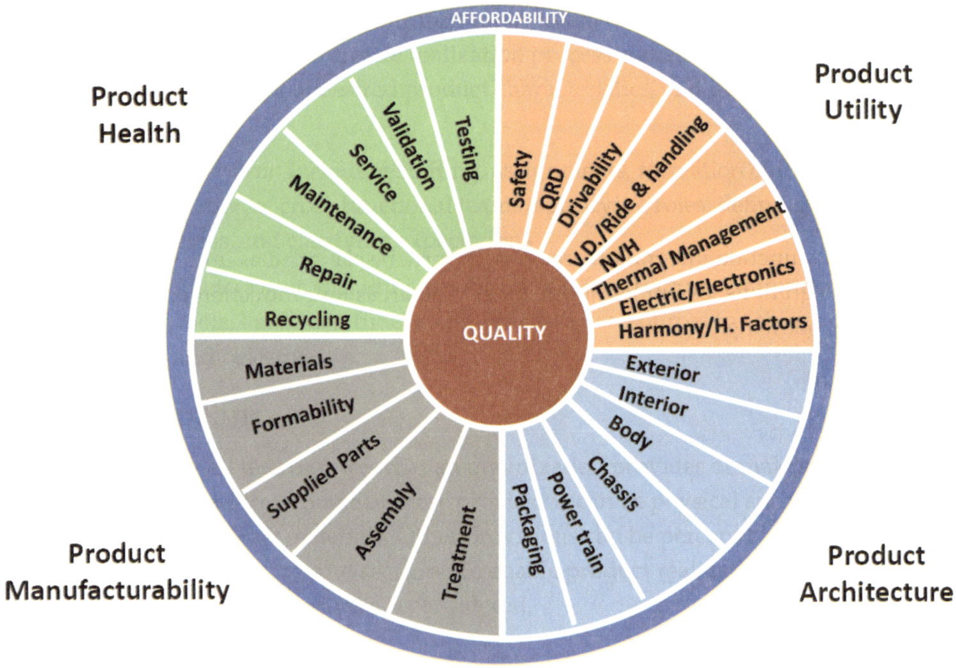

Figure 10.8 Example of different product integration areas.

Some of the product utility activities, shown in Figure 10.8, represent traditional performance integration areas such as safety, QRD, NVH, thermal management, vehicle dynamics, ride and handling, harmony, and human factors.

Simultaneously with product integration activities, production process integration activities must be established to ensure achieving desired process characteristics, attributes, and targets. Considering the production process, form and function characteristics and attributes, the integration areas can be classified as process form integration for form areas, and function integration for performance areas. Integration activities for process fit characteristics, such as quality control and profitability, should be led by the same cross-functional teams and incorporate representations from all process integration areas.

The form integration area can be classified as architectural integration and operation integration. Examples of the integration areas for production process architecture include the following:

- facilities
- equipment
- tooling
- automation

Examples of the integration areas for production process operation include the following:

- safety
- flow
- control
- logistics

Similarly, the function integration area can, in turn, be classified as utility, health, and life cycle integrations. Examples of the integration areas for production process utility include the following:

- capability
- productivity
- adaptability
- consistency

Examples of the integration areas for production process health and life cycle include the following:

- measurement
- cycle time and maintenance
- down time and repair
- scrap and recycling

Figure 10.9 illustrates an example of the different integration areas involved in production process realization.

Each of the integration areas shown in Figure 10.8 and Figure 10.9 has its own structure, participants, and unique activities. Depending on the organizational structure and activities, these integration areas may differ from one organization to another. An example of integration area structure is illustrated in Figure 10.10. The integration area activity is coordinated by the integration area lead. Leads from other involved integration areas, for both product and production process, along with release organizations and suppliers, participate in integration team meetings as needed. Deployment of participating virtual and physical activities is dependent on the realization phase and the integration area needs.

Figure 10.9 Example of different production process integration areas.

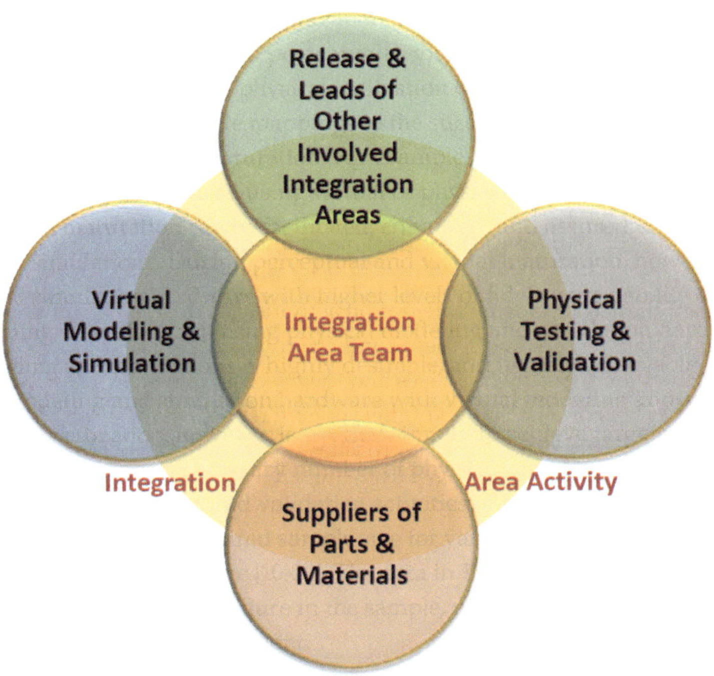

Figure 10.10 Example of integration area structure.

The integration activities of different product areas could be grouped under the major integration areas of architecture, health, utility, and manufacturability, as shown in Figure 10.11. Similarly, the integration activities of different production process areas could be grouped under the major integration areas of architecture, health, utility, and operation, as shown in Figure 10.12. Participation or reporting to the fit integration activities such as quality and quality control could be conducted by these major areas integration leads for both product and process.

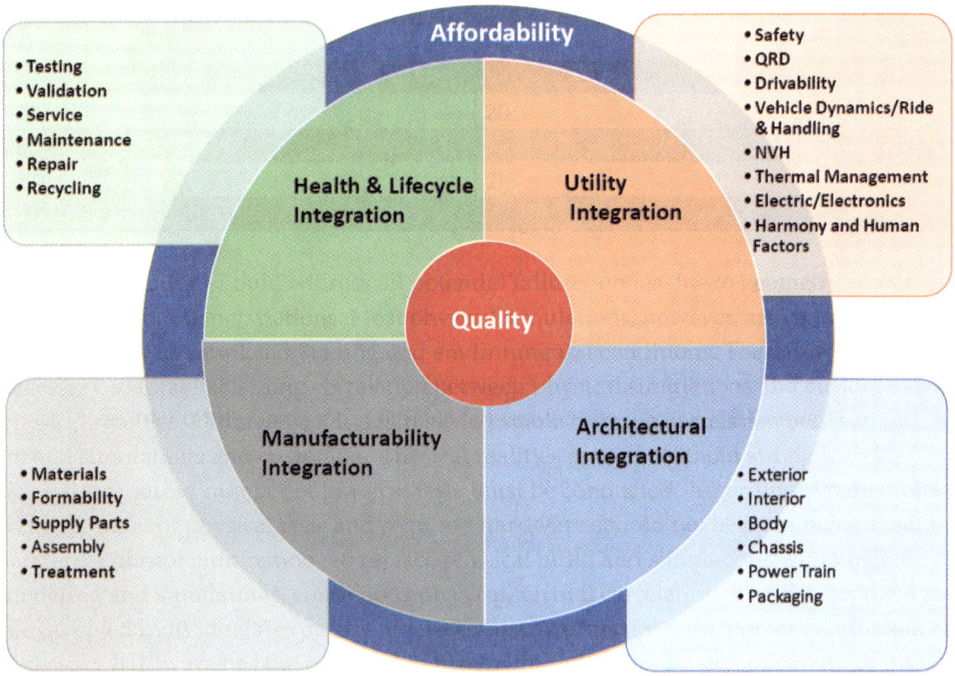

Figure 10.11 Product major integration areas.

One of the key challenges during integration activities is addressing conflicting vehicle targets. Identifying and resolving these conflicts during early realization phases is vital for the success of the subsequent phases and the process. To enable successful resolution of these conflicts or elimination of their potential occurrence, several approaches can be utilized. An example of these approaches is outlined by the flowchart shown in Figure 10.13 [10-11], [10-12], [10-13].

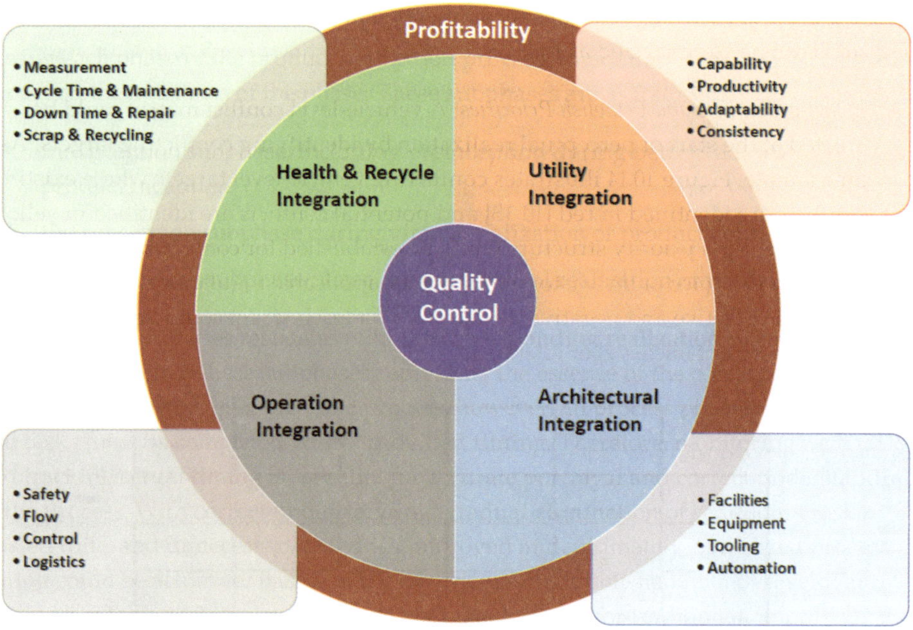

Figure 10.12 Production process major integration areas.

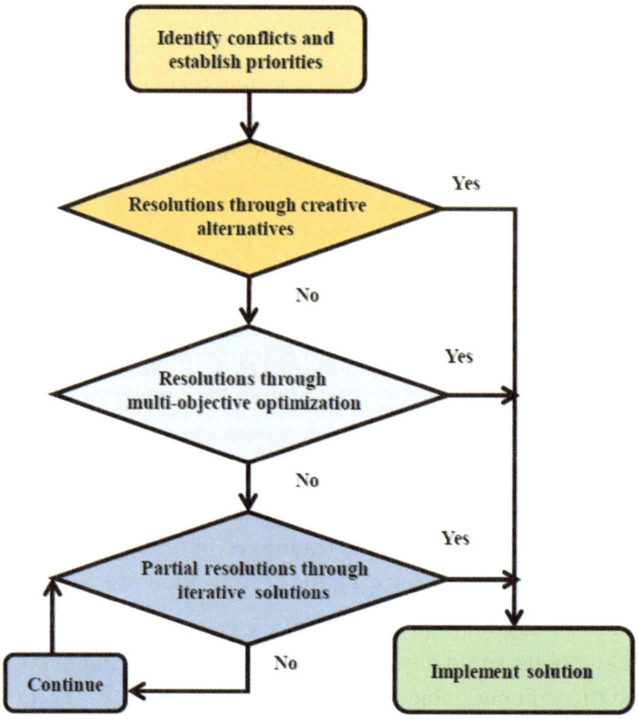

Figure 10.13 The flowchart for a conflict resolution example.

While the color of each step indicates recommended alignment with a realization phase, these steps can be summarized as follows:

1. *Identify Conflicts and Establish Priorities:* A vehicle-level conflict matrix could be created at the start of perceptual realization by identifying conflicting targets. As an example, Figure 10.14 illustrates conflicting vehicle-level targets where existing conflicts are identified in red [10-13] and potential conflicts are identified in yellow. Subsequently, a priority structure could be established for conflicting targets based on the voice of the customer and other applicable regulations. In addition, allowable positive and negative deviations from each target could be specified based on acceptable level of tolerance.

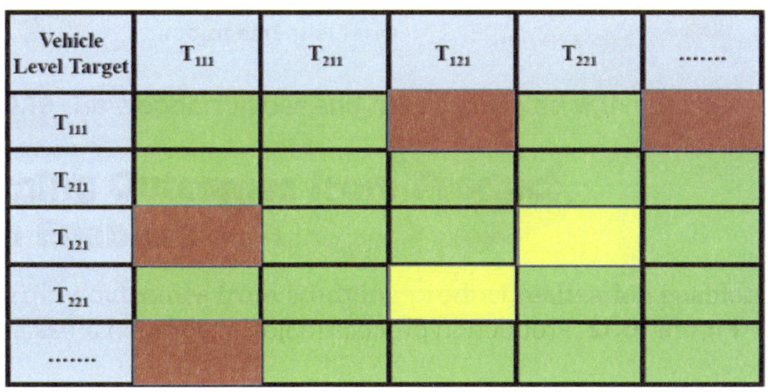

Figure 10.14 An example of targets' existing and potential conflicts.

2. *Resolutions through Creative Alternatives:* Often, there is more freedom to create alternative solutions during early phases of product realization, especially during the perceptual realization phase. Through creative alternatives conflict resolutions could be achieved, and conflicting targets could be decoupled. While it may be attractive to settle for a compromised resolution between conflicting targets, it is more effective to seek creative alternatives that bring synergy or remove the conflict.
3. *Resolutions through Multiobjective Optimization:* Potential resolutions within set priorities could be achieved using multiobjective optimization procedures, if conflict resolution cannot be achieved through creative alternative. As demonstrated by Case Study 10.1, simplified virtual models and simulations could be utilized to formulate the multiobjective optimization procedures. They can also optimize the achievement level of each conflicting target within the established priorities and allowed deviations. Traditional optimization or goal programming tools could be used for performing the optimization procedures [10-10], [10-14].
4. *Partial Resolutions through Iterative Solutions:* Partial resolutions, through alternative iterative solutions, could be attempted if potential resolutions could not be achieved through multiobjective optimization procedures. Through a sequence of design changes and trails using virtual modeling and simulations, an acceptable

alternative that could partially resolve the conflict may be identified. While this procedure is traditionally used, it should be always used as the last resort.

The outlined approach and the steps discussed above could be followed to enable resolution of conflicts during product and production process integration. For example, the architectural conflicts that could be addressed during product integration may include the following:

- architectural conflicts with utility such as styling conflicts with aerodynamics and fuel economy
- architectural conflicts with health such as packaging conflicts with maintainability and reparability
- architectural conflicts with manufacturability such as shape conflicts with formability and assembly

Following steps 1 and 2, identification and resolutions of the key conflicts must be assured by the end of the development and validation iterations of perceptual maturation. Step 3 could also be employed during perceptual maturation, depending on the need and the availability of simplified virtual models or modified virtual models of previous products. If the key conflicts and integration issues are not resolved, then the perceptual concept could be halted during concept initiation approval review until a potential resolution is found. If potential resolutions are not feasible or assured, the idea could be recycled through an idea portfolio. Using more representative and detailed virtual models, steps 3 and 4 could be followed for resolution of the remaining and minor conflicts before prototype build. The steps of the outlined approach could be used to address subsystems and components conflicts. Applicable steps could also be utilized for any arising issues until product launch. To illustrate the application of the outlined approach, the following case study is considered [10-13].

10.3.2.1 Case Study 10.2: Styling and Crushability Conflict: Styling concepts with limited front overhang, as shown in Figure 10.15, usually conflict with crash safety frontal and oblique impact targets.

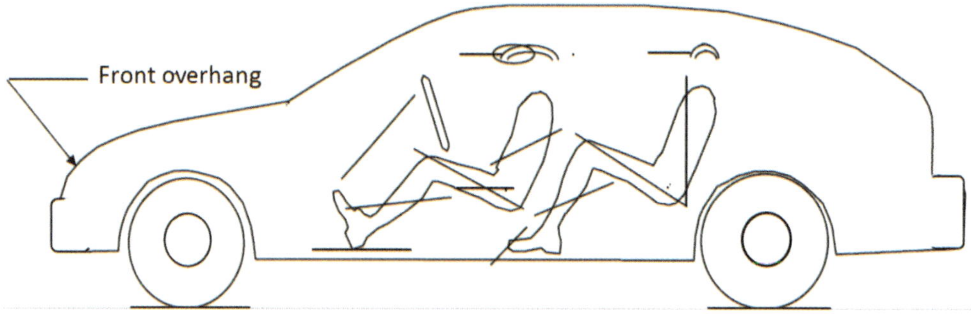

Figure 10.15 The minimum front overhang concept.

During frontal and oblique crashes, longitudinal rails are the main load path and energy management elements. These rails, however, are very difficult to modify after the vehicle architectural concepts and packaging studies are finalized. With the limited front overhang and the resulting limited longitudinal rails crushability, conflicts between styling and crash safety targets must be resolved by the end of the vehicle perceptual realization phase.

Following the approach outlined in Figure 10.13, steps 1 and 2 could be utilized to identify and resolve the conflicts between the crash safety targets and styling through creative longitudinal rails shape and size alternatives. Subsequently, steps 3 and 4 could be used to determine the location and shape of any needed crash initiators, in addition to resolving potential conflicts with durability and NVH.

10.3.3 Management

Management as key organizational activity is the ultimate enabler to product realization and organization continued success. In fact, management is the enabler that initiates, nurtures, and sustains all other enablers, starting with hiring and empowering people, developing and improving processes, and finally securing and controlling provisions.

As part of organizational management, process management enables the achievement of desired product realization outcomes, process timing, affordability, profitability, quality, and customer satisfaction. To enable and assure the success of the product realization process, the quality management system (QMS) model [10-15], [10-16], discussed in Chapter 1, could be modified to account for needed additional customers' reviews and separation of people empowerment from provisions management, as shown in Figure 10.16.

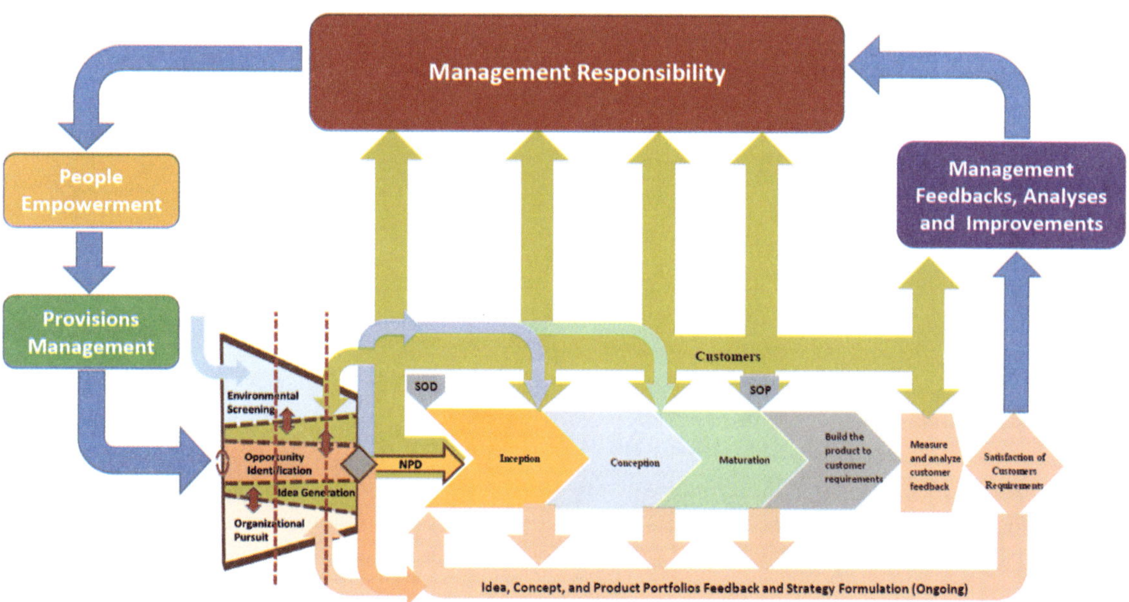

Figure 10.16 The modified quality management system.

Figure 10.16 illustrates the interactions and activities of the following areas:

- *Product Realization:* Including preinception, inception, conception, maturation, product launch, measurement, analysis, and customers' satisfaction feedback
- *Management Feedback, Analyses, and Improvements:* Including product, production process, and realization process feedback, analyses, and proposed improvements
- *Management Responsibility:* Including management obligations, supervisions, decisions, and actions to ensure realization process success, production process control, achievement of desired product characteristics, and ultimately customers' satisfaction
- *People Empowerment:* Including staffing, assignments, formation of integration teams, formation of cross-functional teams, specifying roles, defining responsibilities, and clarifying expectations
- *Provisions Management:* Including provision of infrastructure, equipment, materials, tools, hardware, time, and all necessary resources to achieve desired outcomes and objectives

10.4 Provisions

Provisions include all the resources that an organization provides or makes available for product realization success. Some of these provisions can be physical such as facilities, equipment, tools, and raw materials. Other provisions can be perceptible such as culture, values, encouragement, care, and security. To enable product realization success, the following three critical provisions are considered.

10.4.1 Research and Development

According to KPMG's Global Automotive Executive Survey 2013, the top three most effective growth tactics up to 2018 are [10-17]:

- developing new products and/or new technologies
- entering new markets
- improving affordability

For developing new products and/or new technologies, the top enablers are product and technology research and development (R&D). In fact, product and technology R&D activities are necessary before any new product realization endeavor. However, the maximum impact of product and technology R&D can only be attained through skillful implementation during product realization process, starting from the preinception phase. Unfortunately, product and technology R&D are usually the first activities to lose provisions, support, and finance during hard times and market downturns in many organizations. Some of the main reasons to cutting product and technology R&D are short-term thinking and inability to link R&D to the front end of innovation.

The top enabler for entering or creating new markets is market research. Market research activities are essential to define targeted markets, customers' needs, and the desired product characteristics necessary for successful outcomes before any commitment to product realization. Reducing market research provisions, support, and finance during hard times and market downturns will result only in closing the potential doors leading out of the hard times. Once more, the main reasons to cutting market research are short-term thinking and inability to fully integrate market research findings with the product realization front end.

R&D activities are key enablers for improving total affordability. Continuous waste elimination and cost reduction in R&D are the most effective means to achieve targeted affordability, without sacrificing product quality. To illustrate, the following example is presented.

10.4.1.1 Example 10.1: The major cost components in the production and sale of a vehicle, as categorized by automobile engineering partners, can be summarized as follows [10-18]:

- materials (on average 47% steel, 8% iron, 8% plastic, 7% aluminum, 3% glass, and 27% other materials)
- direct labor
- administration
- R&D
- others (including logistics, advertising, depreciation, etc.)

As shown in Figure 10.17, R&D represents 6% of the cost, while raw materials are 47%. Consider, for example, how a 10% increase in R&D to optimize material usage, find lower cost alternatives, recycle, minimize scrap, and eliminate materials wastes could impact the affordability of the vehicle. Such R&D additional costs would be about 0.6%. Assuming proportional return of 10% reduction in materials costs, a 4.7% cost reduction could be achieved. This would result in a 4.1% overall cost reduction.

While R&D can be conducted for improvement, cost savings, and time reduction, the maximum impact of R&D activities can be achieved when these activities are integrated with the product realization front end of innovation, as discussed in Chapter 6. In addition, these R&D activities should be "focused and tightly aligned with overall corporate strategy" [10-17], [10-19].

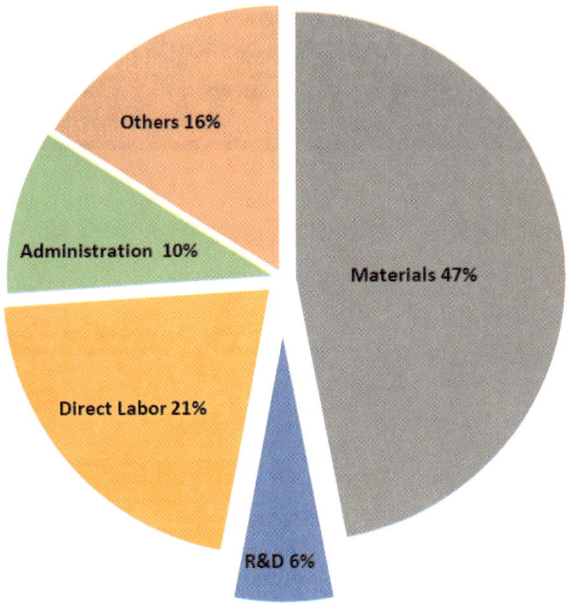

Figure 10.17 Vehicle major cost components.

10.4.2 Hardware

Development and validation during each phase of the product realization process can be performed in the same domain of realization. In other words, perceptual realization can be developed and validated perceptually, virtual realization can be developed and validated virtually, and physical realization can be developed and validated physically. However, using sample mappings to the successive domains could greatly facilitate, enhance, and assure maturation. For example, perceptual maturation can be facilitated, enhanced, and assured using virtual or physical modeling and simulations. Similarly, virtual maturation can be facilitated, enhanced, and assured using physical modeling and simulations. During perceptual and virtual maturation, however, physical modeling and simulation hardware with higher levels of fidelity are usually costly and time-consuming. Therefore, replacing physical modeling and simulation hardware with virtual modeling and simulations is highly desirable, and usually achievable. Replacing all physical modeling and simulation hardware with virtual modeling and simulations during physical realization, however, is not that simple. To achieve targeted reliability and confidence, specific and necessary number of physical samples must be physically tested during the development and validation activities of physical maturation. Table 10.2, for example, shows the required sample size for various population sizes to achieve 90% reliability and 90% confidence [10-20]. The data in Table 10.2 are generated assuming samples without defects and no failure in the sample, using hypergeometric distribution to model sampling without replacement.

Table 10.2 Required sample size for various populations at 90% reliability and 90% confidence [10-20]. (Reprinted with permission from Sandia/John Darby) "sampling without replacement"

Population size	Required sample size
10	9
20	14
30	16
40	17
50	18
70	19
90	20
120	20
150	21
250	21
532 or greater	22

Physical validation should address all potential failure modes due to geometry, materials, and production variations. Most physical simulations, however, are conducted under controlled simulated loading and environmental conditions. Therefore, it is necessary to establish strong correlation between physical simulations and customers' physical realities. Ultimately, it is desirable to establish strong correlation between virtual simulations and customers' physical realities as well. Without strong correlations, the required number of physical tests must be conducted. Accordingly, reduction in the number of physical tests and required hardware should not be attempted through sudden or abrupt elimination. To replace physical build and simulations with virtual modeling and simulations, continuous physical/virtual correlations and development of the involved individuals' expertise are necessary. Any abrupt reduction or elimination of necessary hardware has the potential to turn a limited cost reduction opportunity into uncalculated risk with extreme recalls and warranty costs as illustrated in Figure 10.18.

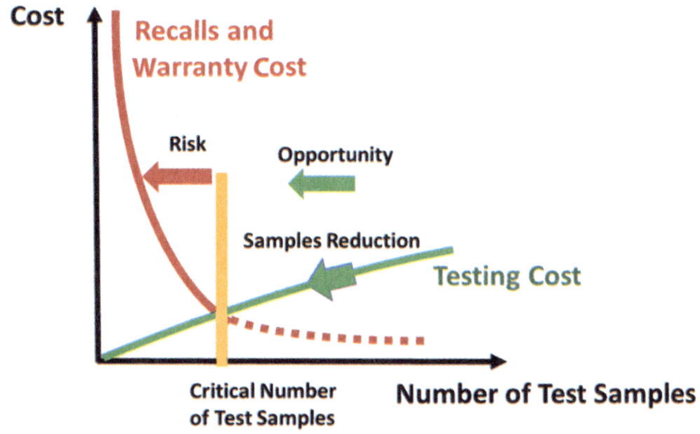

Figure 10.18 Cost opportunities and risks associated with number of test samples reduction.

10.4.3 Time

During each phase of the product realization process, some subphases have significant impact on the outcomes of the phase. These subphases are as follows:

- the inception subphase, including preinception during perceptual realization or product inception
- the conception subphase during virtual realization or product conception
- the maturation subphase phase during physical realization or product maturation

Each of these subphases resonates with the corresponding realization phase. Accordingly, each of these subphases represents the essence of the corresponding phase. To enable on-time completion of the process, however, all process phases, subphases, and tasks must be completed within provided timing. Therefore, monitoring each task and tracking actual timing is essential for accurate portrayal and correct understanding of the process. With correct understanding, gradual elimination of redundant tasks, wasted time, and unnecessary activities improved and, ultimately, optimized process timing could be achieved. Based on the experience and published data of several auto companies, Table 10.3 shows samples of estimated timing for traditional, improved, and optimized vehicle realization processes.

Table 10.3 Examples of estimated timing for traditional, improved, and optimized realization processes.

Process	Duration (months)
Traditional	33–42
Improved	24–33
Optimized	15–24

For process time-reduction, the pressure is always on to eliminate some tasks or reduce the time needed to complete these tasks. Time reduction, however, should not be attempted through abrupt curbing or elimination of required tasks. Without continuous process studies and gradual improvements, reduction of time or elimination of tasks has the potential to turn a limited time reduction opportunity into uncalculated risk with extreme rework and increased number of iterations, as illustrated in Figure 10.19.

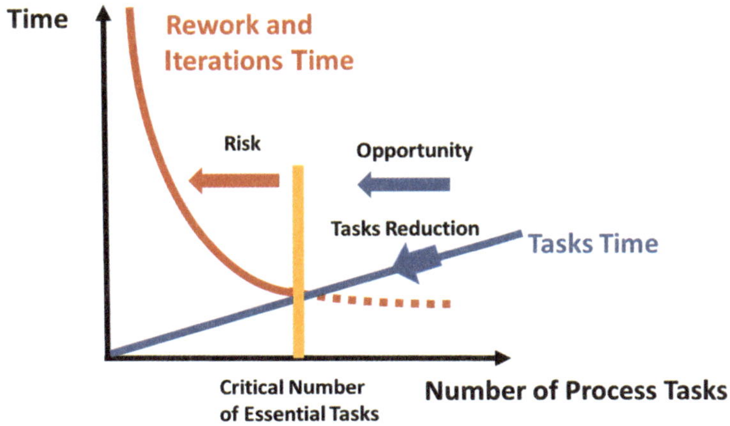

Figure 10.19 Time opportunities and risks associated with reduction of tasks.

10.5 Learning Outcomes from Product Realization Enablers

The desired learning outcomes from studying product realization enablers can be categorized as follows.

10.5.1 Knowledge

People, processes, and provisions are sender- or producer-domain enablers for continued and sustained success of product realization. People in any organization are the main enablers to all activities including product realization. Therefore, continuous development of all people is vital for continued success and growth of the organization. To guide holistic people development, a framework for cognitive, affective, and psychomotor domains, with three progressive realization objectives in each domain, can be established. For the cognitive domain, the three realization objectives or development levels are: understanding, utilizing, and altering. For the affective domain, the realization objectives or development levels are engaging, valuing, and desiring. For the psychomotor domain, these realization objectives or development levels are trying, mastering, and refining. Cognitive and psychomotor development objective can be achieved through education, training, and practice. Affective development objective can be fostered through respect, valuing, and empowering.

Some of the key product realization enablers are connected to the realization process itself. These enablers are associated with analysis, integration, and management. In addition, to the key role of virtual analysis during virtual realization, virtual analysis can guide perceptual conception, perceptual maturation, and enable addressing physical testing and validation issues during physical realization. As for integration, product and production process integration teams and activities must be structured to ensure achieving desired product and production process characteristics, attributes, and targets. Management, however, is the vital enabler that initiates, nurtures, and sustains all other enablers. Management is the sole enabler of people development and empowerment. Management enables realization process, production process, and product development and improvement. In addition, management is the key enabler to securing and controlling necessary provisions for realization process, production process, and product success.

Provisions include all the resources that an organization provides. These provisions include R&D, needed hardware, or allocated process time. While R&D can be conducted for improvement, cost saving and time reduction, the maximum impact of R&D activities can be achieved when these activities are integrated with the front end of innovation. For product and production process success, necessary components, subsystems, and full vehicle physical testing must be conducted and completed during the development and validation activities of physical maturation. Accordingly, sudden or abrupt hardware reduction or elimination could turn an opportunity for cost reduction into a risk with extreme recalls and warranty costs. Similar to hardware, sudden or abrupt time reduction or elimination of tasks could turn a time reduction opportunity into a risk with extreme rework and increased number of iterations.

10.5.2 Skills

Building the right set of skills depends on the individual's role in the organization. For management, the key skill to cultivate is the ability to genuinely make people feel respected, valued, and empowered. Also, the ability to identify when an opportunity may turn into a risk is essential for managing provisions.

For vehicle integration team members, several skills are needed. In addition to the member's specific technical skills, teamwork and communication skills are essential. Also, the ability to identify and resolve conflicts between product or production process targets is necessary for achieving desired attributes.

10.5.3 Paradigm/Attitude

The key to enable successful realization process, product quality, and continued organizational growth is respect for all people. This respect must be extended to include customers, suppliers, stakeholders, society, and eventually everyone starting with the organization internal groups. All employees, independent of their positions and functions, should be valued and empowered instead of being considered or treated as mere resources. Considering some people in the organization as resources divides the organization into decision makers and commodities. Treating people as a commodity will ultimately deteriorate their loyalty, reduce their trust, weaken their initiatives, and dampen their motivation. Conversely, through empowerment, people take ownership, feel appreciated, grow professionally, and ultimately become more productive, creative, and personally fulfilled.

10.6 References

10-1. Bloom, B.S. (Ed.), Engelhart, M.D., Furst, E.J., Hill, W.H., and Krathwohl, D.R. Taxonomy of Educational Objectives: The Classification of Educational Goals. Handbook 1: Cognitive Domain. New York: David McKay. 1956.

10-2. Krathwohl, D.R., Bloom, B.S., and Masia, B.B. 1964. Taxonomy of Educational Objectives: The Classification of Educational Goals. Handbook II: The Affective Domain. New York: David McKay.

10-3. Simpson, B.J. "The Classification of Educational Objectives: Psychomotor Domain," *Illinois Journal of Home Economics* 10(4):110–144. 1966.

10-4. Dave, R.H. 1970. Psychomotor Levels. Developing and writing educational objectives (pp. 33–34). R.J. Armstrong (Ed). Educational Innovators Press: Tucson AZ.

10-5. Bloom, B.S. "Mastery Learning: Theory and Practice." Holt, Rinehart & Winston: New York. 47–63, 1971.

10-6. Harrow, A.J. 1972. A Taxonomy of the Psychomotor Domain: A Guide for Developing Behavioral Objectives. David McKay: New York.

10-7. Simpson E. J. "The Classification of Educational Objectives in the Psychomotor Domain." Gryphon House: Washington, DC. 1972.

10-8. Anderson, L. W., and Krathwohl, D. R. (Eds) 2001. "A Taxonomy for Learning, Teaching and Assessing: A Revision of Bloom's Taxonomy of Educational Objectives: Complete Edition." Longman: New York.

10-9. M. El-Sayed and J. El-Sayed. "Importance of Psychomotor Development for Innovation and Creativity," *International Journal of Process Education* 4: 89–94, 2012.

10-10. El-Sayed, M. and Song, D., "Automotive Performance Optimization," SAE Technical Paper 980825, 1998, doi:10.4271/980825.

10-11. M. El-Sayed, T. Stawiarski, and J. El-Sayed "Multi-Attribute Balancing Process for Automotive Components," ASME Paper No. IMECE2011-65009, *Proceedings of the IMECE 2011 International Mechanical Engineering* Congress, Denver, Colorado, 2011.

10-12. El-Sayed, M. and El-Sayed, J., "Balancing Manufacturability and Performance Attributes in Lean Design," *SAE Int. J. Mater. Manf.* 5(1):174-182, 2012, doi:10.4271/2012-01-0512.

10-13. M. El-Sayed and J. El-Sayed "Resolving Conflicting Product Attributes," Paper#: 30134, ICMAAE 2013 Conference, July 2013.

10-14. El-Sayed, M. E. M., and T. S. Jang. "Structural Optimization Using Unconstrained Nonlinear Goal Programming Algorithm." *Computers and Structures* 52(4): 723–727, 1994.

10-15. Automotive Industry Action Group. ISO/TS 16949:2002 Implementation Guide, CQI-7, Automotive Industry Action Group: Michigan. 2003.

10-16. Hoyle, D., 2009. *ISO 9000 Quality Systems Handbook: Using the standards as a framework for business improvement*. Routledge.

10-17. KPMG's Global Automotive Executive Survey 2013. Available from: https://www.kpmg.com/KZ/ru/IssuesAndInsights/ArticlesAndPublications/Documents/KPMGs-Global-Automotive-Executive-Survey-2013.pdf.

10-18. Kallstrom, H. "Raw materials—the biggest cost driver in the auto industry," Market Realist, 2015. Retrieved January 2016, Available from: http://marketrealist.com/2015/02/raw-materials-biggest-cost-driver-auto-industry/.

10-19. Jaruzelski, B., Loehr, J. and Holman, R., 2011. Why culture is key. *Strategy and Business, 65*(1), pp. 1–17.

10-20. Darby J. L. "Sample Sizes for Confidence Limits for Reliability," Sandia National Laboratories, SAND2010-0550, 2010.

About the Author

Dr. Mohamed El-Sayed, P.E., is the director of the School of Engineering Technology at Eastern Michigan University. For over twenty years, he served as a professor of mechanical engineering and director of the Vehicle Integration & Durability Laboratory at Kettering University in Michigan. He is a well-recognized technical leader in vehicle integration, vehicle development, optimization, and validation. He is an SAE and ASME fellow, as well as the editor in chief of the SAE International *Journal of Materials and Manufacturing*, and the inaugural chair of the SAE Journals' Editorial Board.

In addition, Dr. El-Sayed is a member of the editorial boards of numerous other technical journals. He has served as a board member, advisor, and consultant to several organizations. Through his research, teaching, and industrial practice he has made important original contributions to advance the state of the art in product development, vehicle integration, design optimization, lean design, integrated design and manufacturing, and theoretical and applied mechanics.

Dr. El-Sayed has over thirty years of industrial, teaching, and research experience, several patents granted, and more than a hundred publications in his fields of expertise. He is an award-winning educator, especially in the areas of engineering capstone project courses and online education. Through his teaching and advising, he has contributed to the education of hundreds of engineers now engaged in the field of automotive engineering and product development.

Dr. El-Sayed has also worked as a chief engineer, director of engineering, lead engineer, and subject matter expert in the automotive industry. At General Motors, he led the quality, durability, and reliability integration, development, and validation of several vehicles, platforms, and architectures.

Index

Note: Page numbers followed by *'f'* and *'t'* represent figures and tables respectively.